SOLVING KEPLER'S EQUATION

OVER THREE CENTURIES

SOLVING KEPLER'S EQUATION

OVER THREE CENTURIES

PETER COLWELL

Published by:
Willmann-Bell, Inc.
P.O. Box 35025
Richmond, VA 23235

Published by Willmann-Bell, Inc.
P.O. Box 35025, Richmond, Virginia 23235

Copyright ©1993 by Willmann–Bell, Inc.

First published 1993

Printed in the United States of America

Library of Congress Cataloging-in-Publication Data.

Colwell, Peter.
    Solving Kepler's equation over three centuries / Peter Colwell.
        p.   cm.
    Includes bibliographical references and index.
    ISBN 0-943396-40-9
    1. Kepler's equation –Numerical solutions–History. I. Title.
QB355.5.C65   1993   93-6379
521'.3–dc20   CIP

93 94 95 96 97 98 9 8 7 6 5 4 3 2

Preface

It has been my pleasure to gather and study all I could of the literature of Kepler's Equation. In 1966 Victor Smith, a student who became a friend, asked me what had motivated the astronomer Bessel to describe Bessel functions. Although our friendship has attenuated with time and distance, and it was 1986 before I gave thought to Victor's question, scratching the itch it caused has led to this book.

I express my appreciation to J.M.A. Danby of North Carolina State University, R.H. Gooding of the Royal Aircraft Establishment, and L.G. Taff of the Space Science Telescope Institute for their written work and the attention they gave to mine as it emerged. Also, I owe thanks to Owen Gingerich and Brian G. Marsden of the Harvard-Smithsonian Center for Astrophysics and Jean Meeus who were kind enough to read and comment on my manuscript. The end result is far better for their help, and responsibility for inevitable gaps and errors lies, of course, with me.

P.C.
Ames, Iowa
February 1993

Table of Contents

Preface   iii

Introduction   ix

1 Origins, Antecedents, and Early Developments   **1**
The Anomalies and Kepler's Equation 1
Kepler's Solution 4
Parallax and Arab Encounters with KE 4

2 Nonanalytic Solutions   **7**
Solutions Not Ascribing to Kepler's Second Law 7
The Cassini Solution 12
The Horrocks Solution 16
The Horrebow Solution 18
Solution by Cycloid 20
Solution by "Curve of Sines" 21

3 Infinite Series Solutions   **23**
Solution by Lagrange's Theorem 23
KE and Bessel Functions 27
Levi-Civita's Solution of KE 38
A Lie-Series Solution of KE 41

4 Solutions of KE by Iteration   **45**
Kepler's Solution Revisited 46
Newton's Method and KE 48
Ivory's Geometric Iteration 54

5 Solutions of KE for High Eccentricity   **57**
Barker's Equation and Parabolic Approximations 57
Gauss' Method 61

6 Cauchy and KE **67**
   Cauchy's Treatment of Lagrange's Theorem and KE 69
   Following Cauchy from 1849 to 1941 74

7 Calculations, Auxiliary Tables, and Analogue Devices **79**
   Tables and Approximation Formulas 80
   Analogue Devices 86

8 Modern Treatments of KE **93**
   The Period 1930–1950 93
   "Universal" Forms for KE 98
   Numerical Experiments with KE 98
   KE and Methods for Transcendental Equations 108
      The Burniston-Siewert Method 109
      The Ioakimidis-Papadakis Method 111
      The Delves-Lyness Method 112
      Newton's Method for Power Series 113

A Geometric Parallax and KE **119**

B Error in the Horrocks Solution **121**

C Machin's E_1 **125**

D Coefficients for the Lagrange Solution **127**

E Coefficients of the Levi-Civita Solution **129**

F Autonomous Differential Equations and Lie-Series **135**

G Coefficients of the Lie-Series Solution **137**

H Binary Systems and KE **139**

I Hyperbolas and Battin's Universal KE **141**

J Boltz's Parameters **147**

K Riemann Boundary Value Problems and the Burniston-Siewert Method **149**
   Riemann Boundary Value Problem 149
   The Burniston-Siewert Method and KE 151

L Newton's Method for Formal Power Series **155**

M References Sorted by Categories   159

Bibliography   169

Index   199

Introduction

The sole subject of our work is *Kepler's Equation* (KE)

$$M = E - e \sin E.$$

In its narrowest form, *the Kepler problem* is to solve KE for E, given M in the interval $[0, \pi]$ and e in the interval $[0, 1]$.

In virtually every decade from 1650 to the present there have appeared papers devoted to the Kepler problem and its solution. We can see from a list of them that the problem has enticed a wide variety of scientists to comment on or involve themselves in its solution.

It is surely not unique in science for a specific problem to be given so much attention over so long a period—particularly if it resists solution, if its partial solutions are inadequate or unsatisfactory, or if it is recharged with new interpretations and new applications. Still, it is curious that the Kepler problem should have continued to be this interesting to so many for so long. Admittedly it is a problem central to celestial mechanics, but it is a technical little problem for which a number of satisfactory solutions are long known. With the advent of calculators and computers, there is no impediment to achieving quick solutions of great accuracy. The problem has neither the broad appeal of an Olbers Paradox, nor the depth and intractability of a many-body problem.

In common with almost any scientific problem which achieves a certain longevity and whose literature exceeds a certain critical mass, the Kepler problem has acquired an undeniable luster and allure for the modern practitioner. Any new technique for the treatment of transcendental equations should be applied to this illustrious test case; any new insight, however slight, lets its conceiver join an eminent list of contributors.

The Kepler problem has been "on the scene" in Western civilization science for over three centuries. To gather its story is to view this science through a narrow-band filter, and our goal is to make the picture at one wavelength instructive and interesting.

The idea to gather all the work on the Kepler problem is not new. Early work was surveyed in [Melander,1767], [Detmoldt,1798] and [Brinkley,1803], and much of the recent work was described in [Danby,1983], [Burkhardt,1983] and [Gooding,1985]. In addition, various partial bibliographies have appeared which contain extensive references to the Kepler problem. Perhaps the most cited has been [Radau,1900]; others are [Struve,1860], [Houzeau,1882], [Herglotz,1910], and [Wood,1950a].

Our concern will be almost exclusively for the elliptic case, $0 \le e < 1$, of Kepler's problem. What coverage we attempt of the hyperbolic case, $e > 1$, will appear in Chapter 8 when we look at universal forms of KE, and the parabolic case, $e = 1$, will be considered in Chapter 5, where treatments for high eccentricity (e near 1) cases are gathered. Related to KE and the Kepler problem is the subject of Lambert's theorem and the Lambert orbital problem. Except for a few bibliography entries, we won't describe this area at all, even though it has an extensive literature and independent interest. The references [Gooding,1988,1990] are to be recommended for a look at the subject.

In the language of the mathematician, there exists a unique solution of the Kepler problem, but there are many methods to describe or approximate it. It isn't feasible to say how many there are: the meaning of "solution" depends too much on the solver's motivations. In our classification of references bearing on KE which appear in the Bibliography, we have made seven primary categories and five secondary categories of methods of solution. These have been adequate to describe almost all our references.

Chapter 1

Origins, Antecedents, and Early Developments

After years of work, several false starts, and the need to overcome many personal difficulties, Johannes Kepler announced his three laws of planetary motion early in the seventeenth century. How they and the soon-to-come observations of Galileo changed the thinking of the time and the course of science to follow has been studied and described by philosophers and historians ever since.

Kepler's three laws state:

1. The planets move along elliptical paths with the sun at one focus.

2. As each planet moves about the sun, the line connecting the sun with the planet sweeps out equal areas in equal times.

3. The ratio of the square of the period of revolution of each planet to the cube of the semimajor axis of its orbit is the same for all planets.

Kepler stated the first two laws in 1609 in the *Astronomia Nova* and the third in 1619 in *The Harmony of the World*. As we know, these laws were only approximations, but for the six planets known at the time and to the limits of observation then they were nearly exact.

The Anomalies and Kepler's Equation

In astronomical usage an *anomaly* is an angular quantity used to describe position in the orbit of a celestial body. The description of KE involves three such anomalies.

Suppose a planet moves in the counterclockwise direction in an elliptical orbit about the sun of eccentricity e, $0 < e < 1$, of semimajor axis a, and T represents the time for one complete orbit.

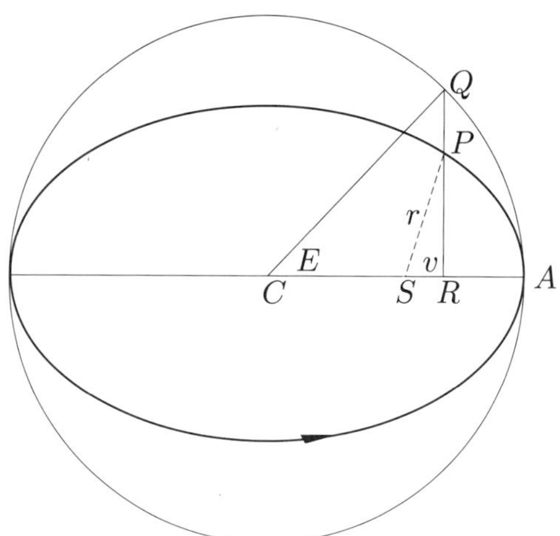

Figure 1.1

If the planet, having passed through perihelion A, is at position P after elapsed time t, we wish to express the polar coordinates of P, (r, v), relative to the sun in terms of t.

The quantity $v = \angle PSA$ is called the *true anomaly* of the planet at time t. If C is the center of the orbit, the eccentric circle is the circle centered at C with radius a. We draw the line through P perpendicular to the radius CA and mark R, its intersection with CA, and Q, its intersection with the circle; the quantity $E = \angle QCA$ is called the *eccentric anomaly* of the planet at time t.

A relation between r and v at time t is given by

$$r = \frac{a(1 - e^2)}{1 + e \cos v}. \tag{1.1}$$

With $b = a\sqrt{1 - e^2}$, it is known that

$$\frac{b}{a} = \frac{PR}{QR} = \frac{\text{Area } PRA}{\text{Area } QRA} \tag{1.2}$$

so that

$$r \sin v = b \sin E \tag{1.3}$$

$$\sin v = \frac{b}{r} \sin E$$

$$r \cos v = CR - CS = a \cos E - ae \qquad (1.4)$$

and

$$r = a(1 - e \cos E). \qquad (1.5)$$

Now

$$\tan \frac{v}{2} = \frac{\sin v}{1 + \cos v} = \frac{r \sin v}{r + r \cos v}$$

$$= \frac{b \sin E}{a(1 - e \cos E) + a(\cos E - e)}$$

$$= \frac{b \sin E}{a(1 - e)(1 + \cos E)}$$

$$= \frac{\sqrt{1 - e^2} \sin E}{(1 - e)(1 + \cos E)}$$

and

$$\tan \frac{v}{2} = \sqrt{\frac{1 + e}{1 - e}} \tan \frac{E}{2}. \qquad (1.6)$$

Thus (r, v) can be determined from E.

KE relates E to time t by means of a quantity

$$M = \frac{2\pi t}{T}$$

where T is the time required for the planet to complete one trip in its orbit about the sun. The quantity M is called the *mean anomaly* of the planet at time t, and it represents the average angular speed of the radius vector SP as it turns through angle 2π in time T.

Kepler's second law implies that

$$\text{Area } PSA = \left(\frac{t}{T}\right) \text{ (Area enclosed by the orbit)}$$

$$= \frac{t\pi a^2 \sqrt{1 - e^2}}{T}$$

$$= \frac{1}{2} M a^2 \sqrt{1 - e^2}.$$

Also

$$\text{Area } PSA = \text{Area } PSR + \text{Area } PRA$$

$$= \frac{1}{2}a(\cos E - e)(r \sin v) + \frac{b}{a}\text{Area } QRA$$

$$= \frac{1}{2}ab(\cos E - e)\sin E + \frac{b}{a}\left(\frac{1}{2}a^2 E - \frac{1}{2}a^2 \sin E \cos E\right)$$

$$= \frac{1}{2}a^2\sqrt{1 - e^2}(E - e\sin E)$$

so

$$M = E - e\sin E. \tag{KE}$$

To summarize, if we know t and M, and if we can solve KE for E, then

$$r = a(1 - e\cos E) \tag{1.5}$$

$$\tan\frac{v}{2} = \sqrt{\frac{1 + e}{1 - e}}\tan\frac{E}{2} \tag{1.6}$$

will determine position (r, v) at time t.

Kepler's Solution

Kepler's own solution of KE went this way: given e and M, guess an approximate solution E_0 for E and calculate

$$M_0 = E_0 - e\sin E_0.$$

Let $E_1 = E_0 + (M - M_0)$ and calculate

$$M_1 = E_1 - e\sin E_1.$$

Then $E_2 = E_1 + (M - M_1)$ should be a better approximation of E.

With the observational accuracy of the time, and for the planets then known, this scheme would produce acceptable values of E in two steps— although it would be somewhat troublesome for Mercury, whose eccentricity is greater than 0.2.

Parallax and Arab Encounters with KE

What we call Kepler's Equation and Kepler's method of solving it appeared in the ninth century writings of Habash-al-Hasib in connection with problems of parallax.

For the planets observable by eye and for stars relatively near the earth, two observers at widely-separated locations on earth will require different coordinates to locate the same celestial object. The parallax problem is that of determining how to transform one viewer's coordinates to the other's, or how to transform both sets of observations to frame of reference standard for all viewers. One commonly-used standard might be coordinates of the object relative to the center of the earth, presumed to be a perfect sphere. So, translating viewers' coordinates in spherical coordinates presents a fine exercise in trigonometry, and we describe details in Appendix A in terms of geocentric parallax so that one can see how a copy of KE emerges naturally from the parallax problem.

History of science of such antiquity is known to require of its scholars more expertise and surmise than history of times where more unambiguous source materials are available. The origin of the method used by Kepler and Habash-al-Hasib is not known, but A.P. Iushkevich, [Iushkevich,1966,1976], and E.S. Kennedy, [Kennedy,1956], think it likely that the mathematics is Hindu rather than Arab.

Neither Habash-al-Hasib nor Kepler established that the iteration scheme for KE must converge to a solution; an argument that this is so may be found in [Kennedy and Transue,1956].

Chapter 2

Nonanalytic Solutions

Kepler's Laws provide a model for planetary motion which crudely fits observational data. Newton's inverse-square law of gravitational attraction leads to a theory which subsumes this model and explains planetary motion. The extent to which Kepler stated and established the Laws as we have stated them, the extent to which astronomers from Kepler's time to Newton's accepted or agreed with these Laws, and the extent to which Newton derived the theory which subsumes them have been the subject of research and conjecture which continues to the present time. Appraisal of this research and contributions to it may be found in [Wilson,1968,1970,1989] and in [Stephenson,1987]. After the work of Lagrange and Laplace, and especially after 1800, Newton's theory had such universal acceptance that virtually all efforts to solve KE were based in analysis.

We use the label "nonanalytic solutions" to denote a varied collection of efforts to describe the position of planets in terms of time. Some of these efforts accepted that the orbits are elliptical but not that they should satisfy Kepler's equal-area principle. In such cases, of course, we are not actually dealing with solutions of KE, but some of the results were influential and fully as accurate as many results obtained from solving KE. Other efforts used models mixing circular and elliptical paths. Many of these solutions are geometric in character, and it is interesting to find them used frequently and rediscovered during the period 1850–1920.

Solutions Not Ascribing to Kepler's Second Law

In 1936 W.C. Brenke published a short paper, [Brenke,1936], recounting that, for an elliptical orbit about the sun, if one plots mean anomaly M relative to perihelion at the "empty" focus, the angle θ relative to perihelion at the sun satisfies the relation

$$\frac{\tan \dfrac{M}{2}}{\tan \dfrac{\theta}{2}} = \frac{1 - e}{1 + e}. \tag{2.1}$$

And for $e < 0.25$, θ is a good approximation to the true anomaly v corresponding to M.

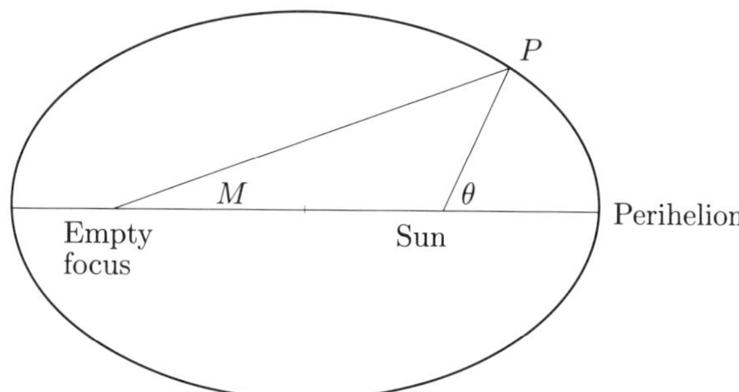

Figure 2.1

What Brenke noticed was first described in 1626 by Albert Curtz, [Gaythorpe,1925], who was among the first to accept Kepler's first law. In lieu of the second law, Curtz assumed that a body moving along an elliptical path about the sun describes equal angles in equal times about the empty focus. This assumption relates M and v by the equation

$$\frac{\tan \dfrac{v}{2}}{\tan \dfrac{M}{2}} = \frac{1 + e}{1 - e} \tag{2.2}$$

and provides a model to determine v from M.

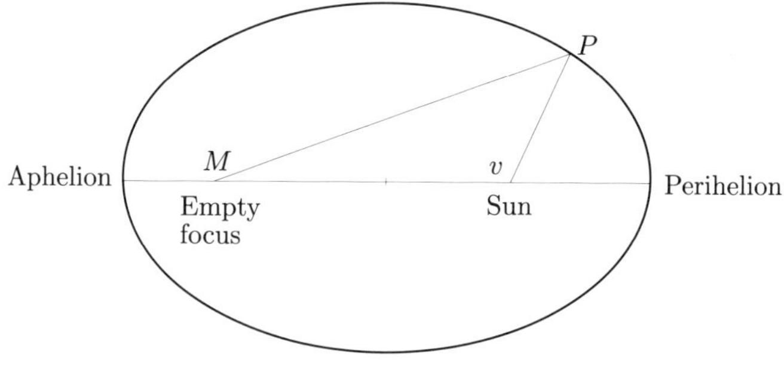

Figure 2.2

The relations (2.1), (2.2) are attributed to William Niele by J. Brinkley, [Brinkley,1803, p. 94], whose survey of the solutions of the Kepler problem until 1800 is our central reference. (The pagination of the Brinkley paper is erratic, and we cite the page numbers as they are printed.) Niele's rule is a consequence of the law of tangents.

Although it was acknowledged that the rate of change of v with respect to time is not constant, the belief that planetary motion should be uniform with respect to M allowed acceptable results and seemed simpler and more appealing than the principle of equal areas claimed in Kepler's second law. Of the solutions we'll describe of Boulliau, Ward, and Mercator, Brinkley, [Brinkley,1803, p. 146], says they

> ... only adopted the orbit, but not the law of motion in orbit...
> No one attempted to compare his own hypothesis with Kepler's discovery of the equable description of areas. Each considered his own hypothesis as resting upon as solid foundations as Kepler.

These solutions described v in terms of M directly and are not actually solutions of Kepler's problem. (In later chapters we'll describe other efforts to obtain v without the intermediate step of solving KE for E.)

Ismael Boulliau's original hypothesis, [Boulliau,1645], was that the elliptical path should be regarded as a section of a cone whose axis passes through the empty focus, and the motion should be such that the line joining the planet to the cone's vertex should display a uniform motion. Seth Ward in 1653 showed that Boulliau's hypothesis is equivalent to the Curtz assumption, and he used Niele's rule to obtain the true anomaly. The value so obtained is usually called "Seth Ward's anomaly." Brinkley, by series methods, calculated its difference from the correct value to be on the order of e^3. The Seth Ward anomaly was also arrived at in 1657 by Comes Paganus, [Jorgensen,1974, p. 177].

Feeling somewhat slighted by Ward's treatment of him and acknowledging that both his and Ward's solutions would not agree with observations for Mars, Boulliau responded in 1657 with a new hypothesis, known as Boulliau's correction of Ward's hypothesis.

Marking the mean anomaly relative to aphelion at the empty focus, constructing a semicircle at the center of the orbit of diameter equal to the major axis, and marking R on the orbit as the point with elevation angle M at the empty focus relative to aphelion, Boulliau let Q be the point of intersection of the semicircle with a line through R perpendicular to its diameter. He assumed the planet, P, would have position at the intersection of the ellipse with a line from Q to the sun.

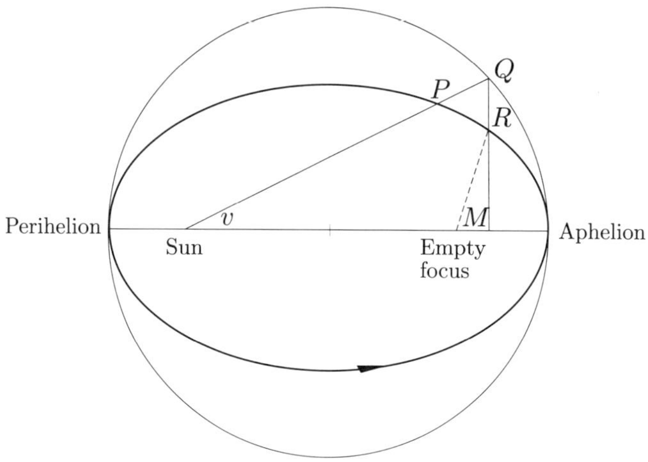

Figure 2.3

Brinkley calculated Boulliau's anomaly—that is, Boulliau's approximation of v under this hypothesis—and found its difference from v to be also on the order of e^3, with the greatest errors occurring for Mercury. Brinkley determined that for small e the ratio

$$\frac{\text{maximum error in Ward's hypothesis}}{\text{maximum error in Boulliau's hypothesis}} \approx \frac{3}{8e}$$

so that for the known planets Boulliau's anomaly is in some sense more accurate than Ward's. Brinkley in 1803 felt he was correcting the record, because the 18th century texts he consulted generally did not regard Boulliau's correction of Ward's hypothesis as an improvement. For the problem of obtaining v, Brinkley wrote, "It cannot, nevertheless, be denied that Ward has obtained more merit than his share in it deserved," [Brinkley,1803, p. 95].

The considerable efforts of S.B. Gaythorpe, [Gaythorpe,1925,1957], in explaining the work of Jeremiah Horrocks on the motion of the moon add perspective to Brinkley's account of Boulliau and Ward. In a letter of 1638, Horrocks considered that the moon followed an elliptical path with the earth at one focus. The details given by Gaythorpe suggest that Horrocks used empty focus constructions similar to those of Curtz. Gaythorpe's findings of Horrocks' original and belatedly-recognized contributions will be described a bit later.

Gaythorpe, in [Gaythorpe,1925], describes the work of Curtz and Horrocks leading to an approximation of v comparable but superior to Boulliau's. Gaythorpe cites a collection of works different from those mentioned by Brink-

11

ley which indicate that Boulliau's approximation was better regarded than Brinkley suggested.

The assumptions of Curtz and Horrocks based on the empty focus construction and the law of tangents will produce an approximation of v with error having an e^3-term equal to $(1/6)e^3\sin^3 M$. The corresponding error for Boulliau's work is $(2/3)e^3\sin^3 M$, greater by a factor of 4. From [Gaythorpe,1920,1957] the Curtz-Horrocks-Boulliau results of the elliptic hypothesis in calculating v can be tracked through a chain of work on KE from 1856 to 1920.

Jean-Dominique Cassini also made a slight contribution relating v to M with a short geometric construction, [Cassini,1669]. Cassini was an eminent astronomer, after whom the "Cassini division" in the rings of Saturn and a hoped-for future space mission to Saturn are named. The reason we have included him here is to avoid confusion: one of the well-known solutions of KE is called "the Cassini solution," but it was the later creation of Jacques Cassini, son of Jean-Dominique. With the exception of the paper of Nicolas Mercator we describe shortly, almost any mention of Cassini in connection with KE one is likely to encounter will refer to [Cassini,1719].

Nicolas Mercator, who is not the Mercator of map-making fame, developed a new hypothesis based on elliptic orbits, [Mercator,1664], and explained it again in a second paper, [Mercator,1670], in response to J.-D. Cassini's paper of 1669. The hypothesis agreed very well with observations of Mars but in the case of no planet did it achieve the accuracy of Boulliau's anomaly, [Brinkley,1803, p. 101].

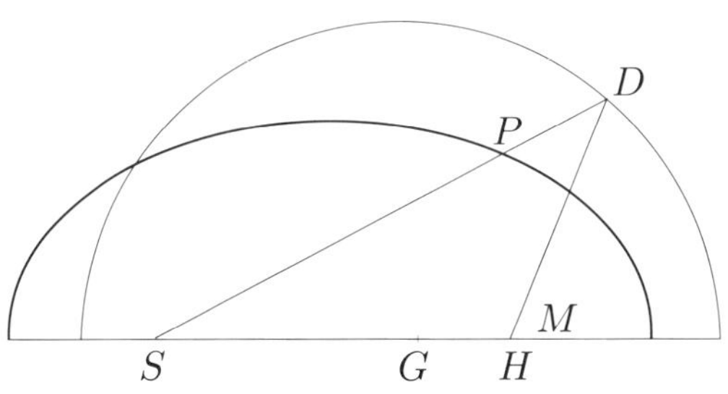

Figure 2.4

If S is the position of the sun, and H is the empty focus, let G be the point between S and H for which $(SG)/(SH) = (GH)/(SG)$—that is, use the golden mean. Describe about G as center a semicircle of radius equal to the orbit's semimajor axis. Choose D on the circle so that $\angle DGH$ is M, the mean anomaly. The angle DSH is Mercator's anomaly. Brinkley calculated that Mercator's anomaly differs from the true anomaly by $-0.014e^2 \sin 2M + (0.548 \sin M - 0.201 \sin 3M)e^3$.

The Cassini Solution

Jacques Cassini, [Cassini,1719], used geometric means to provide a solution of KE which Brinkley, [Brinkley,1803, p. 106], described as

> ... perhaps both the most elegant that has yet been invented, and the most readily deduced from the equable description of areas, and is also, in the planetary orbits, of very great practical value.

Cassini's solution yields a convenient first approximation of E. Cassini himself gave several refinements of it, and many who followed reinvented it, refined it, and embellished it by combining it with other methods. We'll start by describing the method Cassini used and then give several approximations resulting from it and a quick sketch of some of its later manifestations over a period of 150 years.

Let L be the location of a planet with elliptical orbit ALP, perihelion at P, aphelion at A, and sun at S. If C is the center of the orbit, $|AP| = 2$, $|CS| = e$, AJP is a semicircle of radius 1 about C, and

$$\text{mean anomaly } M = \text{arc } AJ = \angle JCA.$$

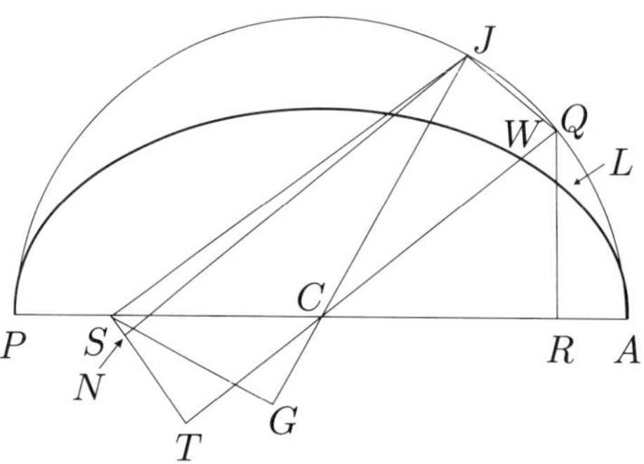

Figure 2.5

The perpendicular to AP through L intersects the semicircle at Q and the diameter AP at R. The remaining lines and points are constructed so that

$$ST \perp QC, \quad SG \perp CJ$$
$$JN \perp ST, \quad JW \perp CQ.$$

The problem is to find

$$\text{eccentric anomaly } E = \text{arc}AQ = \angle QCA$$

since

$$\text{true anomaly } v = \angle LSA$$

is determined by (1.6).

One knows $JN \parallel QT$ and $|NT| = \sin(\angle JCQ)$; from KE it follows that $|ST| = |\text{arc}JQ|$.

From the law of tangents

$$\frac{\tan\left(\dfrac{1}{2}\angle ACJ\right)}{\tan\left[\dfrac{1}{2}(\angle CSJ - \angle CJS)\right]} = \frac{|CJ| + |CS|}{|CJ| - |CS|} = \frac{1+e}{1-e}$$

$$\tan\left[\frac{1}{2}(\angle CSJ - \angle CJS)\right] = \left(\frac{1-e}{1+e}\right)\tan\frac{M}{2}$$

so $\angle CSJ - \angle CJS$ is determined.

Now $\angle CSJ + \angle CJS = M$, so $\angle CSJ$ and $\angle CJS$ are known. Because $E = \angle CSJ + \angle SJN$, we need only $\angle SJN$ to determine E.

Cassini's solution for E is to take $E = \angle CSJ$, assuming $\angle SJN$ is so small as to be negligible. We'll denote this approximation by E_{C}; it is also called Cassini's first approximation.

In fact

$$E_C = \frac{1}{2}M + \frac{1}{2} \text{ (Ward's anomaly)}$$

and Brinkley showed its difference from E is of order e^3.

Cassini took some trouble to argue that, excepting Mercury, $\angle SJN$ is no greater than $5°$ for the other planets and the resulting errors in E_{C} are no greater than 30 seconds of arc. In the case of Mercury, however, $\angle SJN$ is about $24°$ and the error in E_{C} is about 6 minutes of arc.

To improve the approximation, Cassini used

$$\angle SJN \approx \sin(\angle SJN) = \frac{|SN|}{|SJ|}$$

$$= \frac{|ST| - |NT|}{1 + e^2 + 2e\cos M}.$$

Since $|ST| = |\text{arc}JQ| = \angle JCQ$ and $|NT| = |JW| = \sin(\angle JCQ)$,

$$\angle SJN = \frac{\angle JCQ - \sin(\angle JCQ)}{1 + e^2 + 2e\cos M}.$$

Cassini then used an auxiliary table of values of $\theta - \sin\theta$ to calculate $\angle SJN$. Brinkley determined that this improved approximation $\angle CSJ + \angle SJN$ differs from E on the order of e^5.

The Cassini solution, or approximation, for E can be found repeatedly in later literature, often with no recognition that it traces back to Cassini. With reference to Figure 2.5, the essence of the approximation is that line SJ be parallel to line CQ. With reference to the eccentric circle AJP, if one draws a line from the sun's position on the diameter to the point identifying the mean anomaly, a parallel line through the center will intersect the circle at a point approximating the eccentric anomaly.

Exactly this approximation of E appeared in [Hermann,1726], [Stewart,1756], [Ellicott,1799], [Robertson,1816], and [Encke,1850]. In each case the approximation was taken only as a first step and efforts were made for further corrections to obtain a more accurate approximation of E. To give one example of how corrections might proceed, we'll describe the process used by J.F. Encke in [Encke,1850].

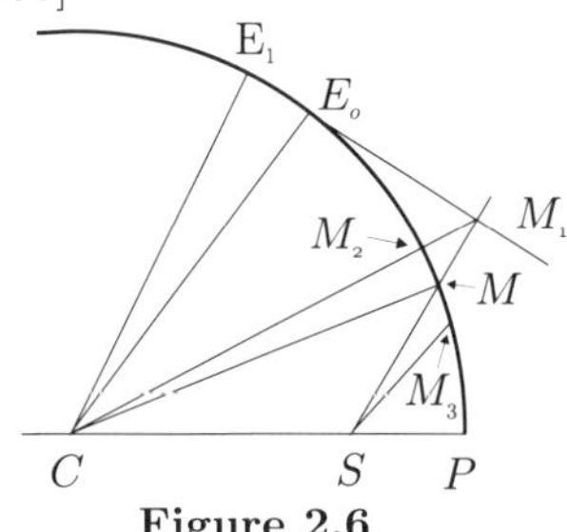

Figure 2.6

On the eccentric circle with center C, perihelion P, and sun's position S, first mark point M so that $\angle MCP$ is the mean anomaly. Choose E_1 on the circle so that CE_1 is parallel to SM—this is the Cassini approximation of E. Let M_1 be the point of intersection of the tangent to the circle at E_1 and the line through S and M. If the line from C to M_1 intersects the circle at M_2, let M_3 be determined on the circle so that

$$\operatorname{arc}(PM_3) = \operatorname{arc}(PM) - \frac{1}{3}\operatorname{arc}(MM_2).$$

Now choose E_o on the circle so that CE_o is parallel to SM_3. The angle $\angle E_\mathrm{o}CP$ is Encke's second approximation of E. If M is the mean anomaly, Encke showed

$$E - E_\mathrm{o} \approx \frac{7}{60}e^5 \sin^5 M.$$

(Recall that $E - E_1$ is of order e^3.) Encke's construction can be continued, with each new approximation improving the accuracy by order e^2. By direct calculation we found that Encke's second approximation with $e < \frac{1}{2}$ finds E accurate to roughly 35 seconds of arc. (In his paper Encke also derived an algorithm based on Newton's method for which the corresponding error is about 1.5 seconds of arc. Newton's method and its association with KE is described in Chapter 4.)

In [Robertson,1816] may be found several different ways of refining Cassini's approximation.

Several analogue devices conceived or actually constructed to solve KE have been based on the Cassini approximation. Many of these are mentioned in Chapter 7, but we'll show in Figure 2.7 a drawing adapted from [Carlini,1856] of a device F. Carlini designed to approximate E.

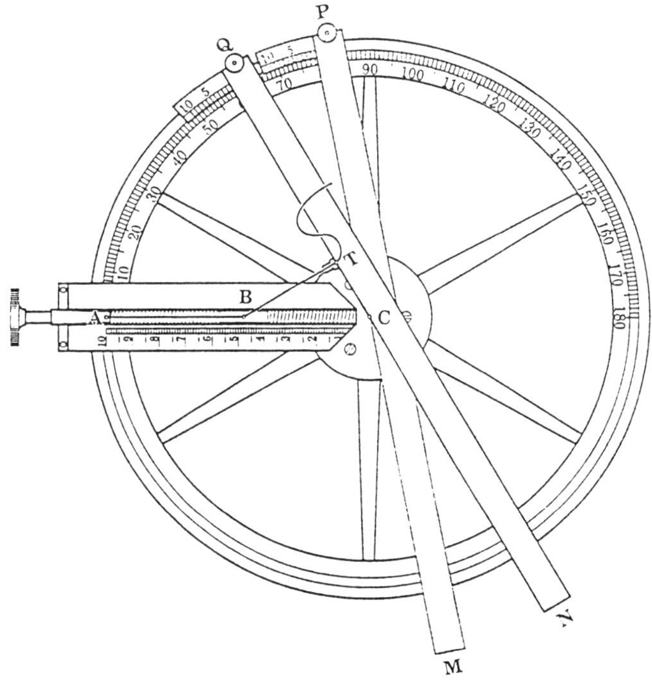

Figure 2.7

In this device the point B is adjusted along a scale to correspond to the value of e, and the movable diameter PCM is set to mark M on the circular scale. Then the diameter is turned with the string ABT being kept taut until the string segment BT is tangent to the inner circle. The position Q on the new diameter position QCN indicates on the circular scale an approximate solution E of the equation $E + e \sin E = M$ under the assumption that $|BT| =$

$|\text{arc } PQ|$. This assumption is in fact equivalent to the assumption in Cassini's approximation. Carlini knew the device can give only a crude approximation of E, and he proposed to use Gauss' method (Chapter 5) to improve it.

The Horrocks Solution

A different approximate method to solve KE is contained in the Jeremiah Horrocks letter of 1638 we mentioned earlier. The provenance of the solution is most interesting, and its identification with Horrocks is the result of relatively recent work by S.B. Gaythorpe. We'll present the solution following [Gaythorpe,1925] and then consider some of its later appearances.

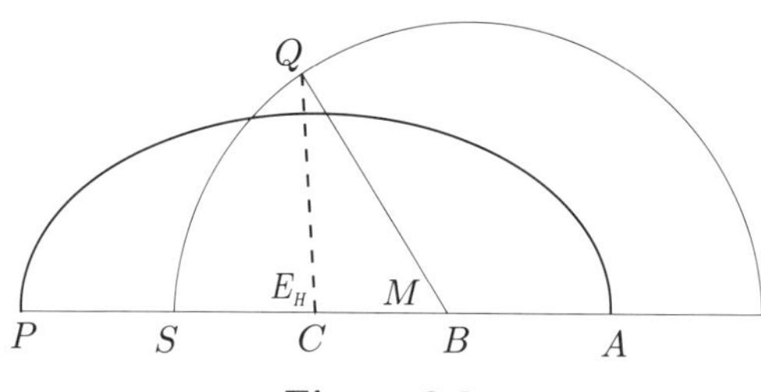

Figure 2.8

About the empty focus B describe a circle of radius PC, the semimajor axis of the elliptical orbit. On this circle mark a point Q so that the mean anomaly $M = \angle QBS$. The Horrocks approximation, E_H, of E is the angle $\angle QCS$.

If PC is normalized to 1, in the triangle QCB

$$\frac{e}{\sin(E_\mathrm{H} - M)} = \frac{1}{\sin(\pi - E_\mathrm{H})} = \frac{1}{\sin E_\mathrm{H}}$$

so

$$\sin(E_\mathrm{H} - M) = e \sin E_\mathrm{H}. \tag{2.3}$$

After dividing the identities

$$\sin E_\mathrm{H} + \sin(E_\mathrm{H} - M) = 2 \sin\left(E_\mathrm{H} - \frac{M}{2}\right) \cos \frac{M}{2}$$

$$\sin E_\mathrm{H} - \sin(E_\mathrm{H} - M) = 2 \cos\left(E_\mathrm{H} - \frac{M}{2}\right) \sin \frac{M}{2}$$

by $\sin E_\mathrm{H}$, there results

$$\frac{1+e}{1-e} = \tan\left(E_H - \frac{M}{2}\right)\cot\frac{M}{2}$$

or

$$\tan\left(E_H - \frac{M}{2}\right) = \left(\frac{1+e}{1-e}\right)\tan\frac{M}{2}. \tag{2.4}$$

With calculations carried to order 3 in e, equation (2.4) implies that

$$E_{\mathrm{H}} - E \approx \frac{1}{6}e^3\sin^3 M. \tag{2.5}$$

(These calculations are made in Appendix B.)

In light of equation (2.3), it is reasonable to replace the problem of finding E in KE with the more tractable problem of finding E_{H} so that

$$E_{\mathrm{H}} - M = e\sin E_{\mathrm{H}} = \sin(E_{\mathrm{H}} - M)$$

or

$$E_{\mathrm{H}} - M = \sin^{-1}(e\sin E_{\mathrm{H}}).$$

Gaythorpe derived equation (2.4) in [Gaythorpe,1920], unaware at the time that it had been used by many investigators to approximate a solution of KE. In the later papers [Gaythorpe,1925,1957], he uncovered the equation's rich background.

Almost at the same time as [Gaythorpe,1920], H.C. Plummer, [Plummer,1919], was ascribing (2.4) to H.A. Howe. The work of Howe to which Plummer was referring appeared in [Howe,1879,1880a,1880b] with slight extension in [Howe,1884]. It is possible that Plummer's attention to Howe's work was caused by the paper [Howe,1919] in which Howe made an appeal for the simplicity and efficacy of his "American solution" of nearly 40 years earlier.

Howe had encountered (2.4) in the work of J.A. Grunert, [Grunert,1856], who used the arcsine series and $E - M = \arcsin\big(\sin(E - M)\big)$ to write

$$e\sin E = E - M = \sin(E - M) + \frac{1}{6}\sin^3(E - M) + \frac{3}{40}\sin^5(E - M) + \dots.$$

Arguing that for small e

$$e\sin E \approx \sin(E - M) \quad\text{and}\quad e \approx \frac{\sin(E - M)}{\sin E},$$

Grunert used about the same identities as Gaythorpe to obtain (2.4). Although in papers earlier than this Grunert showed some acquaintance with astronomical calculations of the 18th century, it still appears that Grunert's discovery of (2.4) was independent. Between 1856 and 1920, equation (2.4) was the basis for many attempts at more accurate solutions of KE. If we let $\Phi = \sin^{-1}(e)$, equation (2.4) may be transformed to

$$\tan\left(E_{\mathrm{H}} - \frac{M}{2}\right) = \tan\left(45° + \frac{\Phi}{2}\right)\tan\frac{M}{2} \tag{2.6}$$

and footprints of (2.4) or (2.6) may be seen in [Radau,1882], [Zenger,1882], [Schering,1884], [Oppolzer,1885], and [Marth,1890a,b]. We have also found (2.4) in [Wallace,1835] with no clue as to its origin, as well as in [Willis,1929].

Equation (2.4) also may be inferred from the details of [Hermann,1726]. The argument for Horrocks' priority, as well as further background on Horrocks and the slight attention he received, may be read in [Gaythorpe,1925].

The Horrebow Solution

Peder Horrebow of Copenhagen obtained a solution in 1717 of KE in the form $M = E + e\sin E$, which is closely related to the solutions of Horrocks, Cassini, and Hermann but not nearly so well known. It is the work of N.T. Jorgensen, [Jorgensen,1974], which brings Horrebow to modern attention and gives some perspective to the details of his solution. These details are described briefly in papers of 1717 and 1718, and again at greater length in a 1742 volume entitled *Adytum Astronomiae,* which resides in manuscript form in the Royal Library of Copenhagen. Jorgensen's paper is our only source in describing Horrebow's work.

In Figure 2.9 let the elliptical orbit be described about center C with the sun at S, empty focus B, and location of the planet at L. Describe circles about C and B of radius equal to the semimajor axis of the orbit, let P be the intersection of the circle about C with the line through L perpendicular to the diameter, and let mean anomaly $M = \angle IBF$.

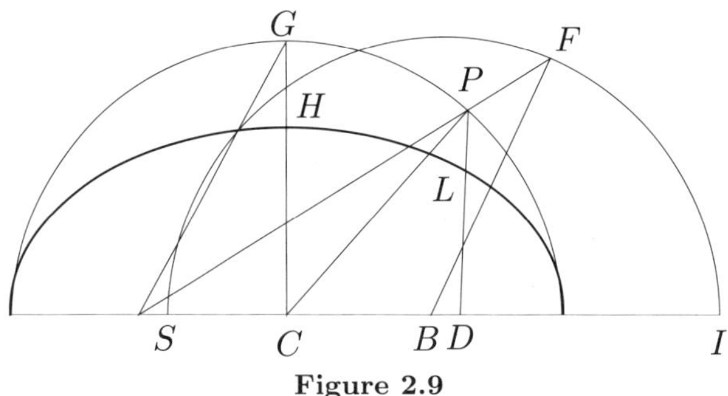

Figure 2.9

Horrebow's first approximation of E is the angle $E_\mathrm{o} = \angle BCF$ and his second is

$$E_1 = \frac{M}{2} + \frac{1}{2}(E_\mathrm{o} - \angle BFC).$$

With the law of tangents on triangle CFB,

$$\frac{1+e}{1-e} = \frac{\tan\dfrac{M}{2}}{\tan\dfrac{1}{2}(E_\mathrm{o} - \angle BFC)} = \frac{\tan\dfrac{M}{2}}{\tan\left(E_1 - \dfrac{M}{2}\right)}$$

and

$$\tan\left(E_1 - \frac{M}{2}\right) = \left(\frac{1-e}{1+e}\right)\tan\frac{M}{2}.$$

It is clear that Horrebow's E_1 is identical to Horrocks' E_H.

The novelty of Horrebow's approach is in the further steps he took to approximate E.

Determining that

$$\text{Area } SPC = [\text{Area } SCG]\sin(\angle BCF)$$

and

$$\text{Area [Sector } APC] = \text{Area [Sector } ASP] - \text{Area } SPC$$

the correct E should satisfy

$$E = M - \text{Area } SPC = M - e\sin(\angle BCF).$$

Since $E_o = \angle BCF$ was corrected to E_1, a better approximation E_2 of E should be

$$E_2 = M - e\sin E_1.$$

In Jorgensen's words, [Jorgensen,1974, p. 176], "Thus Horrebow has reached an iteration process for the determination of E ...." This iteration process is generally credited to Leonhard Euler, [Euler,1740], but it really appears to be Kepler's.

Jorgensen's account draws on Horrebow's work of 1717 and 1742, and between these dates appeared the solutions of Cassini, Hermann, and Euler. Those who haven't studied the sources, as Jorgensen has, can't weigh for themselves Jorgensen's interesting comments on their priority and their later use by others. Jorgensen has shown that Horrebow's work was interesting and relevant. Whether Horrebow knew of Horrocks' work is not established, but Jorgensen has shown that one of Horrebow's students in 1717 knew of Horrocks' 1638 letter.

Solution by Cycloid

A cycloid is a path of motion described by a fixed point on a wheel rolling along a straight line. The cycloid curve holds a large and special place in the history of 17th century mathematics, both because of the interesting questions it posed and because of those it answered in geometry and analysis. For an easy overview it is enough to follow the citations of "cycloid" in the index of either [Kline,1972] or [Boyer,1989].

As we describe it, the cycloid solution of KE is simply a way to use a cycloid to visualize the solution of KE. And, in distinction from all solutions discussed so far, it requires no association of KE with celestial mechanics.

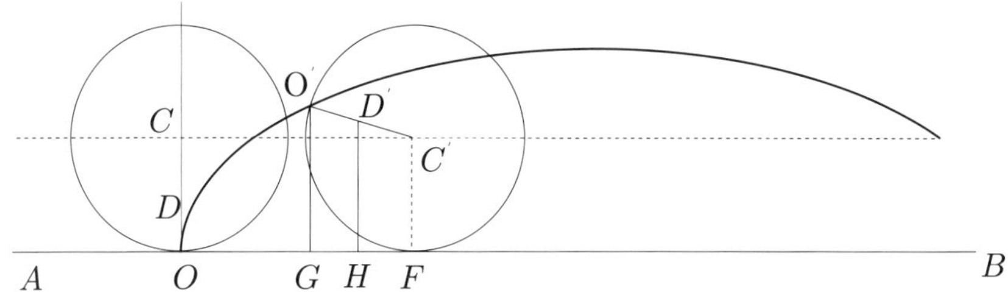

Figure 2.10

Let a circle of radius 1 centered at C be tangent to the horizontal line AB at O. On the line OC let point D be placed so that $|CD| = e$. If the circle now be rolled to the right along AB, the point O will describe a cycloid. If E is the angle through which the line OC has revolved as the circle moves to the point of tangency F, then $|OF| = \angle FC'O' = E$ and the coordinates of O' relative to origin O are

$$(x, y) = (E - \sin E, \; 1 - \cos E).$$

With $O'G \perp AB$ and $D'H \perp AB$, we see $|HF| = e \sin E$.

To solve KE, we roll the circle so that $|OH| = M$; $|HF|$ measures $e \sin E$ and

$$E = M + e \sin E = |OH| + |HF| = |OF|.$$

The foregoing description in the context of the Kepler problem, "... is essentially the solution contrived by Christopher Wren in 1658 and published the next year by John Wallis ...," [Whiteside,1974, pp. 310–311, Note 134]. The solution appears in the section, "De problemate Kepleriano per Cycloidem solvendo," of *De Cycloide Tractatus*, [Wallis,1695]. Isaac Newton in Book I, Proposition 31, of the *Principia* of 1686 used the Wren construction to locate position of a planet in an elliptical orbit; in the Scholium following, he presented a scheme to solve the problem in a more practical manner. (See Chapter 4.)

R.A. Rambaut, H.C. Plummer, and E.J. Wilczynski ([Rambaut,1890], [Plummer,1896], [Wilczynski, 1912]) all reinvented the cycloid solution of KE in the service of planning or building analogue devices to calculate solutions of KE. All were sooner or later apprised of Newton's Proposition 31 ([Rambaut,1913], [Plummer,1907], [Wilczynski,1913]).

In 1901 B. Gonggrijp, [Gonggrijp, 1901], described the cycloid solution, crediting it to Wallis and comparing it with other solutions by construction which had appeared between 1850 and 1900. Gonggrijp assessed for each method the relative value and difficulty of construction, calculation, and approximation.

Solution by "Curve of Sines"

Of the nonanalytic solutions we'll illustrate, solution of KE by the curve of sines is the most recent and perhaps the simplest. Like the cycloid solution, it treats KE with no regard to its origins in celestial mechanics.

In Figure 2.11 we have a graph of $y = \sin x$ in rectangular coordinates. Given M and e, mark point M on the X-axis so that $|OM| = M$. From M draw a line with elevation angle $\Psi = \arctan(1/e)$ relative to the positive X-axis and let P be its point of intersection with the sine curve. If E is the abscissa of P, then P has coordinates $(E, \sin E)$ and $E = M + e \sin E$.

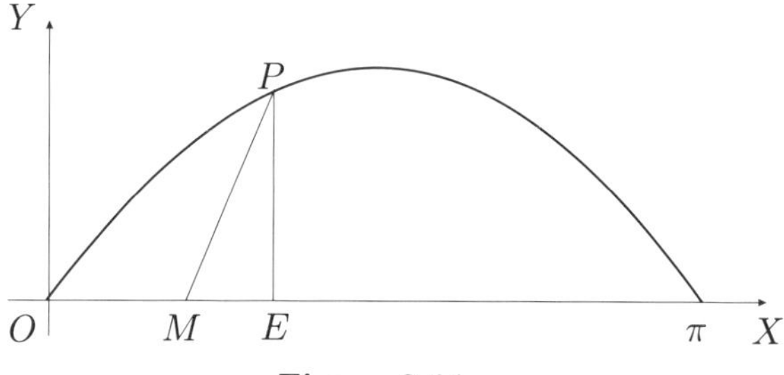

Figure 2.11

As best we can find, this method was the invention of John Couch Adams, [Adams,1849], although he himself said, "The construction is so simple that it has probably been proposed before, though I have nowhere met with it," [Adams,1882, p. 47]. It was reinvented by J.J. Waterston, [Waterston,1850], and E. Dubois, [Dubois,1863], and used by several others in proposing or effecting devices to serve as "KE-solvers," often with some confusion over its origins. (For example, [Proctor,1873], [Pritchard,1877], [Moulton,1895], [See,1895,1896], [Zelbr,1895], and [Dolainski,1913].)

No matter with what care the graph and scales are drawn, the method can give only a first approximation of a solution of KE which most who used it intended to refine by further methods or the use of auxiliary tables.

Chapter 3

Infinite Series Solutions

Of all the solutions known of KE, the infinite series solutions are the most elegant. We'll describe four such solutions: a series in powers of e obtained from a theorem of Lagrange; a Fourier sine series whose coefficients involve Bessel functions of e; a 1904 power series solution of Levi-Civita; and a power series in powers of M whose explicit description is surprisingly recent.

The first of these preceded the work of Cauchy establishing a theory of convergence for power series. The second preceded by nearly a century the work of Dirichlet and Fejer for a similar theory of Fourier series. So, in a sense, both were formal exercises which achieved results far in advance of the theory needed to fully justify them. Both were notable also for the following work they stimulated, both in mathematics and celestial mechanics.

In contrast, the last two series solutions were achieved after the theory they needed was secure. Neither attracted much attention or so far has led to further insight in solving KE. They stand as interesting, but isolated and technical, accomplishments.

Solution by Lagrange's Theorem

Throughout the 18th century, Taylor and Maclaurin series, and power series in general, were used to good effect in the absence of any clear theory of convergence. Sometimes the results were glorious—witness Euler's 1736 argument to show, [Beckmann,1971], [Kline,1983],

$$\sum_{n=1}^{\infty} \frac{1}{n^2} = \frac{\pi^2}{6}.$$

Sometimes the results were puzzling—as when in 1755 Euler decided, [Barbeau,1979],

$$\sum_{n=0}^{\infty}(-1)^n n! = 0.5963474.$$

Many problems of the time were attacked by "reversion of series": to solve equation $y = f(x)$ for x (that is, to find $x = f^{[-1]}(y)$), one would write f in powers of x and then try to invert the series to obtain a power series for $f^{[-1]}$. One of the earliest results to make this feasible was stated by J.L. Lagrange, [Lagrange,1770]:

> Suppose f and ϕ are analytic in a region of the complex z-plane containing a simple closed curve C and its interior. Assume a is inside C and $|t\phi(z)| < |z - a|$ on C. Then the equation
>
> $$w = a + t\phi(w)$$
>
> has a unique solution for w inside C, and

$$f(w) = f(a) + \sum_{n=1}^{\infty} \frac{t^n}{n!} \frac{d^{n-1}}{da^{n-1}} \left\{ f'(a) \left[\phi(a) \right]^n \right\}. \qquad (3.1)$$

From Lagrange's time to the present, this has been a most useful theorem. It was rediscovered by H. Bürmann in 1799, and with refinements provided by Cauchy's calculus of residues it has a modern form called the Lagrange-Bürmann expansion theorem. In modern use it is aimed frequently at problems of scientific computation of the special functions of applied mathematics.

In the special case where $f(z) = z$, $\phi(z) = \sin z$, $a = M$, $t = e$, and $w = E$, the theorem provides an immediate solution of KE in powers of e:

$$\left\{ \begin{aligned} E &= M \sum_{n=1}^{\infty} \frac{e^n}{n!} \frac{d^{n-1}}{dM^{n-1}} [\sin^n M] \\ &= M + \sum_{n=1}^{\infty} a_n(M) e^n. \end{aligned} \right\} \qquad (3.2)$$

Lagrange did not use the theorem on KE in his 1770 paper, but he did so shortly in a 1771 paper devoted to KE, [Lagrange,1771]. In the latter paper Lagrange worked to put this solution more in the form of a Fourier sine series,

$$M + \sum_{n=1}^{\infty} A_n(e) \sin(nM),$$

and we'll say more about his work in this direction when we look at Bessel functions.

This solution by Lagrange's theorem soon achieved a stature as a standard, accessible solution of KE. It was the solution presented by Laplace in his *Mécanique céleste* of 1799, and it became even more entrenched after Bowditch's 1829 translation with extensive commentary. Laplace did calculate the radius of convergence of the series (3.2) and used similar infinite series to represent a variety of related quantities. (Laplace's treatments of convergence of the infinite series in the *Mécanique céleste* later became the foil for Cauchy's establishment of a theory of infinite series, and we'll treat this theory in Chapter 6.)

G. Klugel, [Klugel,1789], although not using Lagrange's theorem, obtained explicit forms for the coefficients

$$a_1(M), \quad a_2(M), \quad \ldots, \quad a_6(M)$$

in (3.2). The operational problem of determining explicit formulas for

$$\frac{d^{n-1}}{dM^{n-1}} \{\sin^n M\}$$

occupied many. F.T. Schubert determined $a_1(M)$ through $a_{11}(M)$ in [Schubert,1818,1820a] and pushed to $a_{13}(M)$ in [Schubert,1820b]. At this point he found discrepancies in his numerical results and those of the then-current literature, leading him to write,

> Perhaps this will move an astronomer to investigate which of us is wrong. I confess, since I have made these time-consuming calculations twice, that I have neither patience nor time to make them a third time... [Schubert,1820b].

C.F. Degen, [Degen,1821], certified Schubert's results, and there were always others willing to have a go at the mind-numbing computational tasks involved. One was C. Creedy, [Creedy,1857], who published many pages of manipulations with results we found simply unreadable. E. Weiss, [Weiss,1884], tried to simplify matters with several changes of variable, but his results are quite inaccurate if $e > 0.3$. R. Bryant, [Bryant,1887], thought to choose f in (3.1) so that $f'\phi^n$ is a constant function for some n. For example, with $n = 2$ and $\phi(z) = \sin z$, the choice $f(z) = \cot z$ renders $f'\phi^2 = -1$, and the resulting expression in (3.1) has no e^2-term:

$$\cot E = \cot M - \frac{e}{\sin M} + \frac{e^3}{6} \sin M + \ldots .$$

However, he found all the cases he tried with $n \geq 3$ to be too complicated.

F.R. Moulton, [Moulton,1907], derived (3.2) without recourse to Lagrange's theorem in this manner: If

$$E = \sum_{n=0}^{\infty} E_n(M)e^n$$

solves KE, then

$$\frac{dE}{dM} = \frac{1}{1 - e\cos E} \tag{3.3}$$

with $E = 0$ when $M = 0$. Arguing that $E_n(0) = 0$ for each n, he rewrote (3.3) in the form

$$\frac{dE}{dM} = \sum_{j=0}^{\infty} (\cos^j E)e^j, \tag{3.4}$$

substituted

$$\sum_{n=0}^{\infty} E_n(M)e^n$$

for E, matched up powers of e, and started solving simple but increasingly recursive differential equations for $E_0(M)$, $E_1(M)$, $E_2(M)$, ... He was able to reproduce some of the coefficients $a_n(M)$ of (3.2). Modern uses of the Lagrange-Bürmann theorem are not so different. In recent years the increased power and accessibility of computer algebra software is making the conceptual simplicity of power series manipulations attractive for solving transcendental equations. Schubert's work to obtain $a_{11}(M)$ in (3.2) can be reproduced in minutes with such software; coefficients through $a_{10}(M)$ appear in Appendix D.

The solution (3.2) was a staple for work in celestial mechanics well into the 20th century. It is, for example, the featured solution of KE in the celestial mechanics textbook of Moulton, [Moulton,1914], which had considerable influence for decades and is cited frequently, even today.

The solution (3.2) in powers of e is convergent only if $e < 0.6627434193$. For planetary orbits where e is very small, this means that values of E can be obtained to acceptable accuracy with relatively few terms of (3.2). For comets, however, e may be too large for (3.2) to be of any use. Halley's comet, whose 1758 return had been predicted by Edmund Halley, is presently assigned $e = 0.9673$. Between 1758 and the next expected return of Halley's comet in 1835, observations of comets and other objects of high eccentricity were of great interest, and the Kepler problem was far from resolved by (3.2).

Laplace, in the *Mécanique céleste*, mentioned that all his series, like (3.2), would converge rapidly but he included no justification. His convergence arguments were made in [Laplace,1823]. There he decided that the series (3.2) converges as long as

$$e < k = \frac{2\sqrt{w(1-w)}}{1-2w} \tag{3.5}$$

where w is the solution of the equation

$$\frac{1-w}{w} = \exp\left(\frac{2}{1-2w}\right). \tag{3.6}$$

Laplace determined $w = 0.08307$ and $k = 0.66195$. As we'll discuss in Chapter 6, Cauchy was not satisfied with Laplace's arguments for convergence, and the correct values for w and k are in fact $w = 0.08322\ldots$ and $k = 0.66274\ldots$. However arrived at, the results (3.5) and (3.6) are correct.

As we'll see in Chapter 5, the Kepler problem for e near 1 was treated quite differently. Whatever success was achieved for e near 1 left a further gap for values of e greater than $0.66274\ldots$ but not near 1. How this gap was filled appears in Chapter 4.

KE and Bessel Functions

With F.W. Bessel's integral definition of what we now call Bessel functions of the first kind of integer order n,

$$J_n(x) = \frac{1}{\pi} \int_0^\pi \cos(x \sin\theta - n\theta)\,d\theta \tag{3.7}$$

in modern notation, there results a very pretty solution of KE in the form of a Fourier sine series:

$$E = M + \sum_{n=1}^\infty \frac{2}{n} J_n(ne) \sin(nM). \tag{3.8}$$

The details of a satisfactory theory of convergence of Fourier series in the early 19th century were even more clouded and remote than had been those of a theory of convergence of power series in the 18th. Fourier's original work on heat diffusion in 1807 implied that continuous functions would be expressible as infinite series of periodic functions. Lagrange, in particular, could not be persuaded that this was sensible, [Grattan-Guinness,1972, p. 24], even though he himself took the first steps to express the solution of KE in the form (3.8) in [Lagrange,1771].

Just as was the situation for power series, the results to be achieved through Fourier series were too compelling to be put on hold while waiting for a theory of convergence—which in the instance of Fourier series didn't arrive until about 1900. Bessel had no qualms in using them to attack some problems in celestial mechanics, including KE. The belief that Bessel invented Bessel functions in order to solve KE is encountered often enough (see, for example, [Gedeon,1964, p. 80]). The actual details are not so simple. After presenting the solution of KE "in the manner of Bessel," we'll describe Bessel's involvement more fully, and then track down some of its consequences between 1820 and 1934 for KE and celestial mechanics. Some discipline and single-mindedness will be required to keep the focus on KE, for there is virtually no corner of applied mathematics after 1820 in which Bessel functions failed to play a substantial part.

If $E = g(M)$ is the solution of KE, g has $M = 0$ and $M = \pi$ as fixed points, and if on $[0, \pi]$

$$g(M) - M = \sum_{n=1}^{\infty} A_n(M) \sin(nM) \tag{3.9}$$

then for $n = 1, 2, 3, \ldots$

$$A_n = \frac{2}{\pi} \int_0^\pi [g(M) - M] \sin nM \, dM$$

$$= -\frac{2}{n\pi} \int_0^\pi [g(M) - M] d(\cos nM)$$

$$= -\frac{2}{n\pi} \{[g(M) - M] \cos nM\} \, |_0^\pi + \frac{2}{n\pi} \int_0^\pi \cos nM [g'(M) - 1] dM$$

$$= \frac{2}{n\pi} \int_0^\pi \cos nM g'(M) dM$$

$$= \frac{2}{n\pi} \int_0^\pi \cos nM \, d(g(M)).$$

Since $M = E - e \sin E = g(M) - e \sin(g(M))$,

$$A_n = \frac{2}{n\pi} \int_0^{2\pi} \cos[ng(M) - ne \sin(g(M))] d(g(M))$$

$$= \frac{2}{n} \left\{ \frac{1}{\pi} \int_0^\pi \cos(nE - ne \sin E) dE \right\}$$

$$= \frac{2}{n} J_n(ne)$$

leading to (3.8).

The functions $J_n(x)$ of (3.7) were not new; in their other now-familiar form

$$J_n(x) = x^n \sum_{k=0}^{\infty} \frac{(-1)^k x^{2k}}{2^{2k+n} k!(k+n)!} \tag{3.10}$$

$$n = 0, 1, 2, 3, \ldots$$

they had been encountered as early as 1703. Euler treated them in considerable generality with his studies of vibrations around 1764, [Whittaker,1927, p. 356]. The connection between Bessel functions and KE dates from [Lagrange,1771]. After obtaining (3.2), Lagrange went far enough in writing E in the form (3.8) to obtain series forms for the coefficients $A_1(e)$, $A_2(e)$, $A_3(e)$.

Bessel was well-acquainted with Lagrange's work. In [Bessel,1805], his first published paper, Bessel worked on the problem of determining the true anomaly, v, from the mean anomaly. This work, which did not involve KE directly and has no hint of his integral definition of Bessel functions, was an effort to improve on the determination of v in high-eccentricity cases where the approach through (3.2) and Lagrange's theorem was known to be inaccurate. This paper of Bessel's will get further treatment in Chapter 5.

In his papers [Bessel,1818b,1819], which are identical, Bessel considered KE and related problems. He considered Lagrange's work to obtain $A_1(e)$, $A_2(e)$, $A_3(e)$, as well as other later efforts to find further coefficients in (3.2), to be crude and clumsy—the word he used was "kunstlos." It was his opinion that Lagrange's theorem was too general a tool for obtaining tractable results in specific instances, and he asserted he had found a more convenient and practical method which would yield cleaner solutions to a variety of problems in celestial mechanics. He chose to display his method on KE as an example, but he didn't consider the Kepler problem to be a most important one.

Instead of Lagrange's procedure to obtain $A_n(e)$ by means of repeated derivatives of products, Bessel cast the computational problems in the form of integrals, much as we now calculate Fourier coefficients. But the results as he left them in [Bessel,1818b,1819] were little improved over Lagrange's, and it is hard to acknowledge that he made much progress in calculating $A_n(e)$.

It is in the landmark paper, [Bessel,1826], that Bessel gave his integral definition of Bessel functions: if h is an integer and k is an arbitrary number,

$$2\pi I_k^h = \int_0^{2\pi} \cos(h\xi - k\sin\xi)d\xi. \tag{3.11}$$

In this paper Bessel displayed the customary infinite series (3.10) for Bessel functions, showed that they satisfy a characteristic differential equation, and derived many now-familiar identities. However, KE does not appear in this paper explicitly, although it is used implicitly in several steps.

In [Bessel,1826] Bessel treated a number of problems of celestial mechanics which all eventually required him to evaluate the integrals

$$\int_0^{2\pi} \cos nM \cos E \, dE \quad \text{and} \quad \int_0^{2\pi} \sin nM \sin E \, dE$$

and which he was able to reduce to evaluations of

$$\int_0^{2\pi} \cos(hE - k \sin E) dE$$

where h is an integer.

The actual solution of KE in the form (3.8) appears in a letter from Bessel to W. Olbers dated 23 April 1818, [Bessel,1818a]. In the letter he expressed his amazement that nobody had yet discovered this solution. In [Burkhardt,1908, p. 96, Footnote 462] it is suggested that Bessel achieved this solution, " ...ohne Kenntnis von seinen Vorgangern zu haben ...," but Bessel's own comments show otherwise.

Histories of mathematics don't linger over Bessel, although his name is invoked adjectivally almost as much as Abel's. Bessel was an intimate of Gauss, and Gauss' letters to Bessel were wide-ranging in mathematics. It is as astronomer and longtime director of the observatory at Königsberg that Bessel is most remembered, and we'd like to make brief mention of two more of his achievements. Bessel is credited with the first reliable estimate of a star's distance from the earth (61 Cygni in 1838), [Roy,1988, p. 11]. In 1844 his observations of Sirius led him to conclude Sirius has a dark, massive companion, which was confirmed in 1863. The spectrum of Sirius B, obtained in 1910, led to the inference that Sirius B must be extremely dense—far too dense to be explained by the then-standard models of star formation. Not until about 1930 were theories of stellar evolution adequate to explain the observations and to classify Sirius B as a white dwarf, [Pagels,1985, pp. 57–58].

Before tracking the later work in KE set in motion from Bessel's, let us describe some work contemporary with Bessel's which gives evidence of other efforts to solve problems in celestial mechanics in the context of Fourier series, and to solve KE specifically in the form

$$M + \sum_{n=1}^{\infty} A_n(e) \sin nM.$$

Much in the way Lagrange determined $A_1(e)$, $A_2(e)$, $A_3(e)$ of (3.9), J.F. Hennert, [Hennert,1804], looked to relate M and v by

$$M = B_0(e)v + \sum_{n=1}^{\infty} B_n(e) \sin nv$$

and got as far as calculating

$$B_0(e) = (1 - e^2)^{\frac{3}{2}} \left[\frac{3}{2}e^2 + \frac{15}{8}e^4 + \frac{35}{16}e^6 \right]$$

$$B_1(e) = (1 - e^2)^{\frac{3}{2}} \left[2e + 3e^2 + \frac{15}{4}e^3 \right]$$

$$B_2(e) = (1 - e^2)^{\frac{3}{2}} \left[\frac{3}{4}e + \frac{5}{4}e^4 + \frac{105}{64}e^6 \right].$$

S.D. Poisson, [Poisson,1811, §221], was very close to Bessel functions in the form (3.7). From

$$r = a(1 - e \cos E) \ \text{ and } \ M = E - e \sin E$$

he wrote

$$r = \sum_{k=0}^{\infty} C_k \cos k(v - M) \ \text{ and } \ M = \sum_{k=1}^{\infty} D_k \sin k(v - M)$$

and determined that

$$C_k = \frac{2a}{\pi} \int_0^{\pi} (1 - e \cos E)^2 \cos(kE - ke \sin E) dE$$

$$D_k = \frac{2}{k\pi} \int_0^{\pi} \left[\sqrt{1 - e^2} - (1 - e \cos E)^2 \right] \left\{ \frac{\cos(kE - ke \sin E)}{1 - e \cos E} \right\} dE.$$

Poisson proposed that these quantities could be calculated by quadrature, and he also wrote the quantities $\cos(kE - ke \sin E)$ in series form, but otherwise he seemed to find little independent interest in the quantities C_k, D_k. That D_k is related to dC_k/de Poisson noticed considerably later, [Watson,1922, p. 6] and [LeFort,1846].

We have no evidence that Poisson actually intended to calculate the quantities C_k and D_k. Curiously, A. Hall, [Hall,1905], took the trouble to express C_k, D_k, $k = 1, 2, 3, 4, 5$, in powers of e as far as e^9, but as far as we are able to find, his efforts sank without a trace. Poisson's coefficients find a much-later echo in the extensive calculations of P. Sconzo, [Sconzo,1933b], who derived

$$v = M + \sqrt{1 - e^2} \sum_{n=1}^{\infty} \frac{H_n(e)}{n} \sin nM \qquad (3.12)$$

where

$$H_n(e) = \frac{1}{\pi} \int_0^{2\pi} \frac{\cos(nx - ne \sin x)}{1 - e \cos x} \, dx.$$

Sconzo's result is an interesting companion to equation (3.8) for the true anomaly. Sconzo worked to rewrite $H_n(e)$ in a form more agreeable for computations, but the results are not good. We have found no citations of Sconzo's work, nor have we encountered (3.12) anywhere else.

In 1817 F. Carlini, [Carlini,1817,1818], used Lagrange's theorem with $a = M$, $t = e$, $w = E$, $\phi(z) = \sin z$, and

$$f(z) = 2 \arctan \left[\sqrt{\frac{1 + e}{1 - e}} \tan \frac{z}{2} \right].$$

Since $f(E) = v$, from (3.1)

$$v = f(M) + \sum_{n=1}^{\infty} \frac{e^n}{n!} \frac{d^{n-1}}{dM^{n-1}} \left\{ f'(M) \sin^n M \right\}. \qquad (3.13)$$

With

$$e = \frac{2\alpha}{1 + \alpha^2}$$

$$f'(M) = \frac{\sqrt{1 - e^2}}{1 - e \cos M} = \frac{1 - \alpha^2}{1 + \alpha^2 - 2\alpha \cos M} = 1 + 2 \sum_{k=1}^{\infty} \alpha^k \cos kM \qquad (3.14)$$

and

$$f(M) = M + 2 \sum_{k=1}^{\infty} \frac{\alpha^k}{k} \sin nM. \qquad (3.15)$$

Carlini used (3.14) and (3.15) in (3.13) with extended and elaborate calculations to determine coefficients $\{G_n(e)\}$ so that

$$v = M + \sum_{n=1}^{\infty} G_n(e) \sin nM \qquad (3.16)$$

and then worked to obtain estimates of $G_n(e)$ when n is large. The quantities $G_n(e)$ are the same as

$$\sqrt{1 - e^2}\,\frac{H_n(e)}{n}$$

from (3.12); in [Taff,1985, p. 59] we find

$$G_n(e) = \frac{2}{n} J_n(ne) + \sum_{m=1}^{\infty} \alpha^m \left[J_{n-m}(ne) + J_{n+m}(ne) \right]$$

and the difference

$$G_n(e) - \frac{2}{n} J_n(ne)$$

has a finite bound independent of n, so the asymptotic behavior of $G_n(e)$ and $(2/n)J_n(ne)$ are the same for large n.

Carlini's work attracted little attention for some time, but in 1850 C.G.J. Jacobi thought it important enough to present a translation with extensive corrections and commentary which filled four numbers of the *Astronomische Nachrichten,* [Jacobi,1850]. The descriptive and computational problems of elliptic motion are formidable, and progress with them required extensive development of and facility with the identities satisfied by Bessel functions. In the context of celestial mechanics, the 1850's saw prodigious efforts to grapple with Fourier-Bessel series expansions of quantities like $r^m \cos n v$ in terms of M, E, or e. P.A. Hansen wrote a string of papers, [Hansen,1852b,c, 1853a,c], culminating in a 100-page effort, [Hansen,1855], to detail these expansions. W. Scheibner took up the asymptotic behavior of J_n for large n in [Scheibner,1856b] after examining in [Scheibner,1856a] the coefficients c_k^n in

$$\left(\frac{a}{r} \right)^n = c_0^n + 2 \sum_{k=1}^{\infty} c_k^n \sin kM.$$

(When $n = -1$, for example, $c_k^{-1} = -(e/k)J_k'(ke)$.)

The mathematics available was considerably enlarged over that used in the 1820's. Cauchy's theory of residues was in full use by this time, and Laurent's theorem of 1843 described the generating function for the sequence $\{J_n(t)\}$,

$$\exp\left[\frac{t}{2}\left(z - \frac{1}{z} \right) \right] = \sum_{n=-\infty}^{\infty} J_n(t) z^n$$

and converted Bessel's integral definition (3.7) to

$$J_n(t) = \frac{1}{2\pi i} \oint_C z^{-n-1} \exp\left[\frac{t}{2}\left(z - \frac{1}{z} \right) \right] dz \qquad (3.17)$$

where C is a simple closed curve in the z-plane enclosing the origin. The representation (3.17) had been anticipated by S. Greathead, [Greathead,1837], who started with (3.1) in the manner of Carlini and derived an expression equivalent to (3.16):

$$v - M = 2 \sum_{n=1}^{\infty} \left\{ \alpha^n \exp\left[\frac{ne}{2}(\alpha^{-1} - \alpha) \right] \right.$$
$$\left. + \alpha^{-1} \exp\left[\frac{-ne}{2}(\alpha^{-1} - \alpha) \right] \right\} \frac{\sin nM}{n}.$$

This result was derived also by A. Cayley, [Cayley,1858], and used to prepare a set of tables in 1861 with which to calculate v and related quantities for different values of e.

J. Bourget, [Bourget,1861], started with the Cauchy numbers,

$$N_{-m,n,p} = \frac{1}{2\pi} \int_0^{2\pi} z^{-m} \left(z + \frac{1}{z} \right)^n \left(z - \frac{1}{z} \right)^p du$$

where $z = \exp(iu)$, and m, n, p are integers, which Cauchy had defined in [Cauchy,1841]. Bourget arrived at (3.8) by quite a different path.

The substitution $z = \exp(iu)$ transforms (3.17) to

$$J_m(ne) = \frac{1}{2\pi} \int_0^{2\pi} z^{-m} \exp\left[\frac{ne}{2} \left(z - \frac{1}{z} \right) \right] du \qquad (3.18)$$

and Bourget introduced the notation

$$(j,n)_m = \frac{1}{2\pi} \int_0^{2\pi} z^{-m} \left(z + \frac{1}{z} \right)^j \exp\left[\frac{ne}{2} \left(z - \frac{1}{z} \right) \right] du.$$

(Notice $(0, n)_m = J_m(ne)$.)

He derived

$$(j,n)_m = \sum_{p=0}^{\infty} \frac{\left(\frac{ne}{2} \right)^p}{p!} N_{-m,j,p}$$

with the special case

$$J_m(ne) = \sum_{p=0}^{\infty} \frac{\left(\frac{ne}{2} \right)^p}{p!} N_{-m,0,p}.$$

Cauchy had expressed the solution of KE in the form

$$E = M + \sum_{\substack{n = -\infty \\ n \neq 0}}^{\infty} K_n \exp(inT)$$

and had written, [Cauchy,1841],

$$K_n = \frac{1}{2\pi} \int_0^{2\pi} z^{-n} e \sin u \left[1 - \frac{e}{2} \left(z + \frac{1}{z} \right) \right] \exp \left[\frac{ne}{2} \left(z - \frac{1}{z} \right) \right] du.$$

Bourget showed

$$K_n = \frac{e}{2ni}(1,n)_n = \frac{1}{in}(0,n)_n$$

with $K_{-n} = -K_n$, so that

$$E = M + \sum_{n=1}^{\infty} \frac{2}{n}(0,n)_n \sin nM$$

which is identical with (3.8); he also wrote

$$E = M + \sum_{n=1}^{\infty} \frac{2}{n} \sin nM \left\{ \sum_{p=0}^{\infty} \frac{(\frac{ne}{2})^p}{p!} N_{-n,0,p} \right\}.$$

The work we have described for the period 1850–1861 was motivated directly by celestial mechanics, but it shows results less sensitive to the needs of astronomers and perhaps less accessible to them. After 1860 the development and refinement of what we now call the special functions of applied mathematics became more and more the business of those whose strongest allegiance was to mathematics. While the involvement of mathematicians in solving KE was to continue to grow, astronomers after 1860 were generally more interested in being able to solve KE quickly and repeatedly, but with only modest accuracy.

From 1860 to some time after 1890, the work on KE was mostly derivative and computational. One exception was the work of H. Gylden, [Gylden,1875a], reviewed in [Gylden,1875b], which expressed v in terms of Jacobi's elliptic functions. J.W.L. Glaisher, [Glaisher,1877a], tried to suggest some approximations which might make Gylden's results more practical, but neither work had much impact. (Glaisher, a prolific but not well-known mathematician of wide interests, played an indirect minor role in the 1888 founding of the New York Mathematical Society, which became the American Mathematical Society. While not dating the beginning of American mathematics, the Society's formation was a benchmark of its coming of age. See [Jackson,1988] for more details.)

Serious work on Bessel functions continued, both for mathematics and celestial mechanics. M.W. Kapteyn, [Kapteyn,1893], studied the possibility of representing functions of some generality by series of the form

$$f(x) = \sum_{n=0}^{\infty} a_n(f) J_n(nx)$$

which are now called *Kapteyn series,* and most of the standard quantities associated with orbits can be so expressed.

In later years Bessel functions were generalized in several ways. The only aspect of these generalizations we choose to include in our study are those which continued some connection to KE.

For example, M.B. Jekhowsky considered a multivariable Bessel function

$$J_q(t_1, t_2, ..., t_n) = \frac{1}{\pi} \int_0^{\pi} \cos\left[qx - \sum_{j=1}^{n} t_j \sin jx \right] dx$$

(not original with him) and in [Jekhowsky,1918] applied it to a "multivariable Kepler's equation"

$$M = E - \sum_{j=1}^{n} e_j \sin jE$$

$$|e_1| + |e_2| + \cdots + |e_n| < 1.$$

For

$$\tan \frac{v}{2} = \sqrt{\frac{1+e}{1-e}} \tan \frac{E}{2}$$

$$\beta = 1 - e^2$$

$$y_j = (-1)^j \frac{2}{j} \frac{e^j}{(1+\sqrt{\beta})^j}$$

$$j = 1, 2, 3, \ldots, n$$

he wrote

$$E = f - \sum_{j=1}^{n} y_j \sin jf$$

and obtained

$$E - M = 2e \sum_{k=1}^{\infty} \left\{ J_{k-1}(kz_1, \ldots, kz_n) \right.$$

$$\left. -\frac{1}{k} \frac{\partial J_k(kz_1, \ldots, kz_n)}{\partial z_1} \right\} \sin kf$$

where

$$z_j = \frac{2e^j}{j(1 + \sqrt{\beta})}$$

$$j = 1, 2, \ldots, n.$$

Another example, with a similar multivariable flavor, is Y. Hagihara's application of two-variable residue calculus to treat a pair of generalized Kepler equations

$$M_1 = E_1 - e_1 \phi_1(E_1)$$
$$M_2 = E_2 - e_2 \phi_2(E_2)$$

when ϕ_1, ϕ_2 are 2π-periodic functions, [Hagihara,1936]. Hagihara used changes of variables, Laurent's theorem, and Cauchy integral formulas, together with (3.17), to write double Fourier-Bessel coefficients of functions of (E_1, E_2) in terms of (M_1, M_2) or (E_1, E_2). Such calculations were more common then in celestial mechanics than they are now.

We'll close our discussion of Bessel functions with mention of a paper of J.O. Fleckenstein, [Fleckenstein,1941], which describes a sequence of operational steps leading from (3.2) to (3.8) to demonstrate that

$$\sum_{n=1}^{\infty} (-1)^n \frac{1}{n!} \frac{d^{n-1}}{dM^{n-1}} (e \sin M)^n = \sum_{n=1}^{\infty} \frac{2}{n} J_n(ne) \sin nM.$$

While the left hand sum converges only for $0 \leq e < 0.6627\ldots$, Fleckenstein used the Fourier expansion theorem to justify that the right hand sum converges for $0 \leq e < 1$.

While (3.2) was used seriously to obtain approximate values of E, the representation (3.8) of E was used more for analytical than computational purposes. Calculating values of $J_n(ne)$ from either (3.7) or (3.10) in order to obtain E in (3.8) would always have seemed impractical.

Levi-Civita's Solution of KE

In 1904 T. Levi-Civita, [Levi-Civita,1904a,b], presented an interesting, elegant infinite series solution of KE. Levi-Civita made many contributions to celestial mechanics; he is most widely-remembered in mathematical circles for his later work in tensor geometry, some of which is standard in treatment of general relativity. His work on KE was purely analytical, and the solution he described had little value for computations and attracted almost no later attention. Aside from a one-line mention in [Moulton,1907], we have seen this solution treated only in [Wintner,1941], where it is derived without reference to Levi-Civita, and in [Kaminskii,1982], where recursive relations for the series coefficients of Levi-Civita's solution are derived.

Levi-Civita proved that if

$$z = \frac{e \exp(\sqrt{1-e^2})}{1 + \sqrt{1-e^2}}, \quad 0 \le e < 1,$$

then KE has solution

$$E = \sum_{n=0}^{\infty} L_n(M) z^n \tag{3.19}$$

convergent for $0 \le z < 1$.

Levi-Civita started by considering e as a complex variable. In order that KE have a unique solution for E as a function of e for all real values of M, solutions of the critical point equation, $1 - e \cos E = 0$, must be avoided, and Levi-Civita examined the system

$$\left\{ \begin{array}{l} E - e \sin E = M \\ 1 - e \cos E = 0 \end{array} \right\} \tag{3.20}$$

to determine the simply-connected region in the e-plane containing $e = 0$ in which (3.20) has a unique solution for E.

If $\cos E = 1/e$, $0 \le e < 1$, then Levi-Civita's variable z, as we see, arises naturally from the system (3.20),

$$\sin E = \frac{i\sqrt{1-e^2}}{e}$$

$$\exp(iE) = \cos E + i \sin E = \frac{e}{1 + \sqrt{1-e^2}}$$

$$\exp(iM) = \exp[i(E - e \sin E)] = \frac{e \exp(\sqrt{1-e^2})}{1 + \sqrt{1-e^2}} = z$$

and Levi-Civita's own treatment of KE made no mention of Bessel functions or the close association the quantity

$$z(e) = \frac{e\exp(\sqrt{1-e^2})}{1+\sqrt{1-e^2}}$$

is known to have with Bessel functions and the solution (3.8) of KE. In [Kapteyn,1893], M.W. Kapteyn had derived a sequence of inequalities, now called the *Kapteyn inequalities*,

$$|J_n(ne)| \leq \left| \frac{e^n \exp(n\sqrt{1-e^2})}{(1+\sqrt{1-e^2})^n} \right| \tag{3.21}$$

$$n = 0, 1, 2, \ldots$$

to aid in the description of the behavior of J_n for large n. Levi-Civita's z is Kapteyn's estimate for $|J_1(e)|$.

Also, Levi-Civita's argument for the convergence of (3.19) made no reference to Bessel functions or the inequalities (3.21). In contrast, Wintner's argument for (3.19), while similar to Levi-Civita's, appeals to the Kapteyn inequalities, [Wintner,1941, pp. 218–222].

For real $e, 0 < e < 1, z(e)$ is an increasing function with $e < z(e) < z(1)$. For pure imaginary $e = ix$,

$$|z(ix)| = \frac{x\exp(\sqrt{1+x^2})}{1+\sqrt{1+x^2}}$$

is increasing in x with $z(0) = 0$ and

$$\lim_{x \to +\infty} |z(ix)| = +\infty.$$

There is a unique number R for which $|z(iR)| = 1$, and in fact $R = 0.66274\ldots$, the radius of convergence of (3.2).

If $e = \rho\exp(iy)$, where $\rho \geq 0$ and y is real, the conditions

$$|\, z(P\exp(iy))\,| = 1$$

$$|\, z(\rho\exp(iy))\,| < 1 \ \text{ for } \ \rho < P$$

define $P = P(y)$ as a unique function of y which has the properties:

$$P(y) = P(-y) = P(\pi + y);$$

P is decreasing on the interval $\left[0, \dfrac{\pi}{2}\right]$;

$$P\left(\frac{\pi}{2}\right) = R;$$

$$P(y) \to 1 \text{ as } y \to 0.$$

If $\gamma = \{e : e = P(y)\exp(iy), y\,\text{real}\}$ in the complex e-plane, then $|z(e)| = 1$ when e is on γ and $|z(e)| < 1$ when e is inside γ. Furthermore, KE has a unique solution for $E(e)$, normalized by $E = M$ when $e = 0$, for each e in the interior of γ.

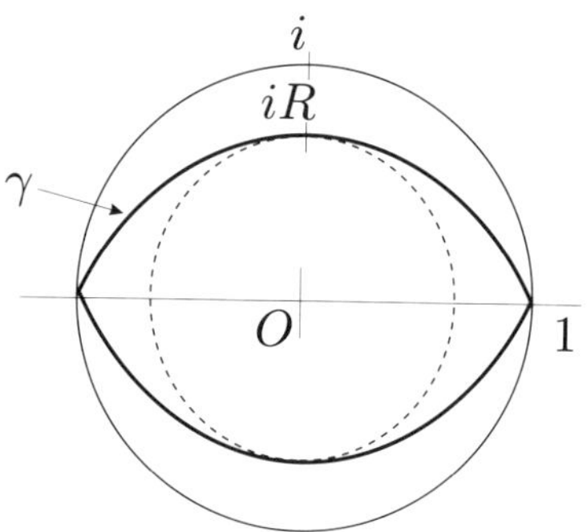

Figure 3.1

This solution is the analytic continuation of solution (3.2), whose region of convergence is the interior of the dotted circle $\{|e| = R\}$ in Figure 3.1.

Since $z(e)$ maps the interior of γ one-to-one and conformally onto the open unit disk $\{|z| < 1\}$ with $z(0) = 0$, if $e = e(z)$ is the inverse mapping, the composite function $E(e(z))$ is analytic in $\{|z| < 1\}$ and so has a Maclaurin series of the form (3.19) for each real M.

The task of determining the coefficients $\{L_n(M)\}$ explicitly in (3.19) is even more formidable than that of determining

$$\frac{d^{n-1}}{dM^{n-1}}(\sin^n M)$$

associated with series (3.2). B. Kaminskii, [Kaminskii,1982], obtained recursive descriptions for these coefficients, but they do not seem suitable for actual calculations. We are not aware that Levi-Civita attempted to write any of the coefficients explicitly or to determine any numerical values of E from (3.19).

In Appendix E we display the coefficients $\{L_n(M) : n = 1, 2, \ldots, 10\}$ as functions of M. Our calculations with the approximation

$$E^* = \sum_{n=1}^{10} L_n(M)z^n$$

of E show that the Levi-Civita solution of KE is not a desirable choice for calculations.

A Lie-Series Solution of KE

If $\mathcal{F}$ is a class of analytic functions with a common domain (which may be complex n-space), and $D : \mathcal{F} \to \mathcal{F}$ is a linear operator on $\mathcal{F}$, *the Lie-series of a function in $\mathcal{F}$ with respect to operator D is*

$$L(D, z, t)(f) = \exp(tD)(f(z)) = \sum_{n=0}^{\infty} \frac{t^n}{n!} D^n(f(z)) \qquad (3.22)$$

where

$$D^0(f(z)) = f(z)$$
$$D^n(f(z)) = D[D^{n-1}(f(z))].$$

The Lie-series of an operator is a sophisticated generalization of the exponential matrix of a square matrix. To those who encounter $n \times n$ systems of autonomous, first-order, ordinary differential equations and who know how useful the exponential matrix is for the representation of solutions of the system, the relevance of Lie-series to more general systems of autonomous differential equations will seem reasonable.

For those whose intuitions have not been altered by encounters with such systems, the technical atmosphere of our discussion has become very chilly and unfriendly. For the price of some patience we hope to describe how the notion of a Lie-series was applied recently to KE to obtain an infinite series solution. The surprise is that, of the small amount of literature connected with this solution, the most important part is relatively new. Our discussion will get to this solution as gracefully and quickly as possible, with technical details for those interested appearing in the Appendices F and G.

The use of Lie-series for systems of differential equations was explicated by W. Grobner in *Die Lie-Reihen und ihre Anwendungen*, Springer-Verlag, Berlin, 1960. Since many interesting problems, including some from celestial mechanics, may be modeled by systems of autonomous differential equations, Lie-series representations of solutions offer both power and elegance in trying to treat various problems consistently.

K. Stumpff, [Stumpff,1968b], illustrated the application of Lie-series to some problems of celestial mechanics, one of them being KE. Stumpff was applying a tool of considerable sophistication to a task not requiring it, but the result was a rational development of an infinite series solution of KE which had received very little attention. After describing Stumpff's procedure, we'll gather what history we've found for the result.

For a single autonomous differential equation

$$\frac{dz}{dt} = F(z)$$

subject to the initial condition $z(0) = \xi$, the method of Lie-series indicates that with operator

$$D(\ ) = F(z)\frac{d}{dz}(\ )$$

the solution will be

$$z(t) = \exp(tD)\xi.$$

For fixed e, $0 \le e < 1$,

$$M = E - e\sin E = f(E) \tag{KE}$$

has the unique solution $E = f^{-1}(M)$ with $f^{-1}(M_0) = E_0$ as long as $f'(E_0) = 1 - e\cos E_0 \ne 0$. And E must satisfy

$$\frac{dE}{dM} = \frac{1}{f'(E)} = \frac{1}{1 - e\cos E}$$

$$E = E_0 \text{ when } M = M_0.$$

If

$$D(\ ) = \frac{1}{1 - e\cos x}\frac{d}{dx}(\ )$$

$$E = \exp[(M - M_0)D(x)|_{x=E_0}] = \sum_{n=0}^{\infty}\left\{\frac{(M - M_0)^n}{n!}D^n(x)|_{x=E_0}\right\}. \tag{3.23}$$

The quantities

$$D^n(x)\,|_{x=E_0}$$

$$n = 0, 1, 2, \ldots \tag{3.24}$$

are difficult to calculate, and the radius of convergence of series (3.23) is quite small unless e is close to 0. However, for M close to M_0, the approximation of (3.23) with just a few terms can be quite good. The series requires us to know $E_0 = f^{-1}(M_0)$, and the values (3.24) must be recalculated if the initial values change.

In the special case $M_0 = 0$, $E_0 = 0$, this entails calculation at $x = 0$ of

$$D(x) = \frac{1}{1 - e \cos x}$$

$$D^2(x) = \frac{1}{1 - e \cos x} \frac{d}{dx} \left(\frac{1}{1 - e \cos x} \right)$$

$$D^3(x) = \frac{1}{1 - e \cos x} \frac{d}{dx} \left(D^2(x) \right)$$
$$\cdots$$

Through $n = 3$ we approximate (3.23) by

$$E \approx \frac{M}{1 - e} - \frac{M^3 e}{3!(1 - e)^4} + \frac{M^5 e(9e + 1)}{5!(1 - e)^7}. \tag{3.25}$$

In Appendix G we display $D^n(x)|_{x=0}$ for $n = 1, 2, \ldots, 15$ and offer some calculations of

$$E_S = \sum_{n=1}^{15} \frac{M^n}{n!} \left\{ D^n(x)|_{x=0} \right\}$$

to compare with accurate values of E obtained by Newton's method. It is hard to agree with Stumpff's assessment of (3.23) as, "... one of the simplest special series involving Kepler's equation ..." [Stumpff,1968b, p. 15].

While from a mathematical viewpoint it might seem as natural to represent E in powers of M as to do so in powers of e, a representation in powers of M has received little attention. The only early reference we have found is [Rittenhouse,1799], where a formula much like (3.25) appears. We have seen a similar expression also in [DeGasparis,1879a]; in neither instance was there any detail to explain where the approximation came from. F.R. Moulton, [Moulton,1903], derived quite carefully information for the region of convergence of a series of form (3.23), but he made no effort to display its coefficients or to use his series to obtain values of E. Moulton's results were derived again by E. Paul, [Paul,1963], also with no display of coefficients. Stumpff's 1968 treatment of the series caused no particular attention as far as we know.

Our treatment of infinite series solution of KE now ends, having spanned both early and recent times. In Chapter 8, in what we think is a novel treatment, we'll revive an old method of Newton for formal power series to obtain in a uniform manner the series (3.2) of Lagrange, (3.19) of Levi-Civita, and (3.23) of Stumpff.

Chapter 4

Solutions of KE by Iteration

A sequence of real numbers

$$\{r_n\}_{n=1}^{\infty}$$

is said to be *defined by iteration* if there is a function of one variable, f, for which

$$r_{n+1} = f(r_n), n = 1, 2, 3, \ldots . \tag{4.1}$$

If such a sequence converges to a limit R, then necessarily $R = f(R)$ and R is called a *fixed point* of f.

An iterative sequence (4.1) is sure to converge to a fixed point of f if the numbers $\{r_n\}$ lie in a closed interval I and there is a constant $L, 0 < L < 1$, for which

$$|f(x) - f(y)| \leq L|x - y| \tag{4.2}$$

whenever x and y are in I.

To solve an equation $F(x) = 0$ for x by iteration is to find a function f and an iterative sequence

$$\{x_n \colon x_{n+1} = f(x_n)\}_{n=0}^{\infty}$$

which converges to a fixed point X of f such that $F(X) = 0$. There is always the choice $f(x) = x - F(x)$ available, and in applied mathematics there are many familiar ways of selecting f. How one decides which f to select depends on F and whatever experience one has had in attempting the solution of equations by iteration. For a given F there may be several feasible choices for f, and one hopes to select in a way that makes $|x_n - X|$ small with a small value of n.

When e and M are specified, solving KE corresponds to solving $F(E) = E - e \sin E - M = 0$ for E. Of all the methods commonly used to solve equations by iteration, two have a strong historical attachment to KE. One is Kepler's own solution as described in Chapter 1, and the other is the celebrated Newton's method.

In this chapter we'll describe also a geometric method of iteration used by J. Ivory which has some interesting features not associated with any other methods we've found. Ivory considered it to have practical value, but we have found no later references to it.

Kepler's Solution Revisited

As we described it earlier, Kepler's solution of KE is equivalent to solving

$$F(E) = E - e \sin E - M = 0$$

by iteration with the "naive" choice

$$f(E) = E - F(E) = M + e \sin E. \tag{4.3}$$

Since in this case

$$|f(E_1) - f(E_2)| = |e \sin E_1 - e \sin E_2| \leq e|E_1 - E_2|$$

the sequence

$$\{E_{n+1} = f(E_n)\}_{n=0}^{\infty} \quad \text{with} \quad E_0 = M$$

will converge if $0 \leq M \leq \pi$ and $0 < e < 1$. For values of e larger than $^1/_2$, many iterations may be required to achieve an accurate approximate solution. (For example, if $M = 0.2$ and $e = 0.75$, $|E_n - f(E_n)| < 5 \times 10^{-4}$ for the first time when $n = 14$.)

Kepler's intention, with small values of e presumed, was to calculate only one or two values E_n; presumably this was the case also for Habash-al-Hasib and the parallax problem.

Leonhard Euler presented this solution in a paper of 1740, [Euler,1740]. While Euler made no effort to prove that the method does converge to a solution of KE, he was accustomed to making extensive, even enormous, numerical calculations and was quite aware that the method is slow when e is near 1.

In 1750, without reference to Euler, N.L. De la Caille, [De la Caille,1750a], attempted to calculate true anomaly v from M for the earth by iterating in the same manner with many added calculations.

With M known, De la Caille started with an approximation v_0 of v and calculated a corresponding value E_0 of E from formula (1.6). From KE he let

$M_0 = E_0 - e \sin E_0$ and proposed $v_1 = v_0 + (M - M_0)$ as a next approximation of v. Repetitions of the loop

$$v_n = \underset{(\text{via } (1.6))}{\longrightarrow} E_n \underset{(\text{KE})}{\longrightarrow} M_n \to v_{n+1} = v_n + (M - M_n)$$

until $M - M_n$ is suitably small completed the scheme. It was recognized by Brinkley, [Brinkley,1803], that with starting value $v_0 = M$ the approximation v_n of v after n steps would have error on the order of e^n.

J.A. Grunert, [Grunert,1851], encountered De la Caille's work and saw it contained a simple iterative scheme to solve KE but seemed not to be aware that it had been described earlier by Euler and Kepler.

The Kepler solution was revived once again by C.F.W. Peters in 1891, [Peters, 1891,a,b]. Peters' work was one of several efforts made between 1860 and 1900 to find ways of solving KE quickly and accurately enough that astronomers might make repeated calculations for determining orbits. These efforts are detailed further in Chapter 7.

Given e and M, Peters suggested taking E_0 from one of several extensive tables of solutions of KE, calculating

$$E_1 = M + e \sin E_0$$
$$E_2 = M + e \sin E_1$$

and using

$$E \approx E_2 - \frac{(E_2 - E_1)^2}{E_2 - 2E_1 + E_0} = \frac{E_2 E_0 - E_1^2}{E_2 - 2E_1 + E_0} \tag{4.4}$$

as the solution of KE.

Modern practitioners of numerical equation-solving would recognize this last step as the essence of Aitken's method for accelerated convergence:

$$\text{if } \{x_n\}_{n=0}^{\infty} \text{ converges to } X,$$

with suitable hypotheses

$$\left\{ y_n = x_n - \frac{(x_{n+1} - x_n)^2}{x_{n+2} - 2x_{n+1} + x_n} \right\}_{n=0}^{\infty} \text{ also converges to } X$$

$$\text{and } \lim_{n \to \infty} \left(\frac{y_n - X}{x_n - X} \right) = 0.$$

Peters ascribed the step (4.4) to Gauss and Oppolzer and concluded his papers by suggesting further approximations, if required, be made by setting

$$E_{n+1} = M + e \sin E_n$$

which is precisely (4.3).

Newton's Method and KE

Newton's method, or the Newton-Raphson method, attempts the solving of an equation $F(x) = 0$ by starting from an initial guess x_0 to generate an iteration sequence with

$$f(x) = x - \frac{F(x)}{F'(x)}.$$

If f and F are suitably nice,

$$|f(x) - f(y)| = |f'(c)||x - y|$$
$$= \left| \frac{F(c)}{F'(c)} \times \frac{F''(c)}{F'(c)} \right| |x - y|$$

for some value of c between x and y. To satisfy criterion (4.2) one hopes that

$$\left| \frac{F F''}{(F')^2} \right| < 1$$

for all points in an interval covering the solutions sought. With reasonable cooperation from F and an initial guess x_0 close to a true solution X, the convergence of the sequence

$$\{x_n \colon x_{n+1} = f(x_n)\}_{n=0}^{\infty}$$

to X can be very rapid. When F does not satisfy these ideal conditions, iterations of f can be very erratic. Many refinements and embellishments of the method are known to extend its utility and relative simplicity to less-than-ideal cases. To our knowledge, none of these refinements has direct historical connection with KE.

The connection of Newton's method with KE is very strong, however. Of the extensive material available on Newton's method, we'll only sample carefully to concentrate on KE. Particularly important sources are [Brinkley,1803], [Adams, 1882], F. Cajori's 1911 account of Raphson's development of the method, Cajori's revision of Motte's 1729 translation of Newton's *Principia,* [Newton,1934], and Volume 6 of D.T. Whiteside's monumental study of Newton's mathematical papers, [Whiteside,1974].

The first edition of the *Principia* appeared in 1686. The third appeared in 1726, and Cajori's revision mentioned above is of the 1729 Motte translation of this third edition. Tracking the description of Newton's method applied to KE through the various editions is difficult, even with the expertise of Cajori and Whiteside to help.

As we would apply it today to KE, Newton's method would:

let E_0 be an initial approximation of E;
define for $n = 0, 1, 2, \ldots$

$$E_{n+1} = E_n - \frac{(E_n - e \sin E_n - M)}{1 - e \cos E_n}.$$

Each E_{n+1} should approximate E more closely than E_n.

J.C. Adams, [Adams,1882, p. 45], wrote that the method

> ... explained above is equivalent to that given by Newton in the *Principia* at pp. 101, 102 of the second edition, and at pp. 109, 110 of the third edition, when Newton's expressions are put into the modern analytical form.

According to Cajori, [Cajori,1911], Newton described a method of approximation of roots in 1669 in *"De analysi per aequationes numero terminoraum infinites,"* published in 1704, and in *"Methodus fluxionum et serierum infinitorum"* in 1671 (published in 1736). In both tracts Newton used the equation

$$x^3 - 2x - 5 = 0$$

as an example. This equation has a solution a bit larger than 2. Here is a look at Newton's calculations.

Assume $X = 2 + p$ is the solution near 2.

$$(2 + p)^3 - 2(2 + p) - 5 = 0$$

$$p^3 + 6p^2 + 10p - 1 = 0. \tag{4.5}$$

Ignoring all but the linear part of (4.5),

$$10p - 1 = 0$$

so let $p = 0.1 + q$ in (4.5):

$$(0.1 + q)^3 + 6(0.1 + q)^2 + 10(0.1 + q) - 1 = 0$$

$$q^3 + 6.3q^2 + 11.23q + 0.061 = 0. \tag{4.6}$$

From the linear part of (4.6),

$$11.23q + 0.061 = 0$$

let

$$q = \frac{-0.061}{11.23} + r = -0.0054 + r$$

in (4.6). Setting the linear part of the resulting equation for r equal to zero and solving produces $r = -0.00004854$.

To this stage

$$X = 2 + p + q + r = 2 + 0.1 - 0.0054 - 0.00004854$$

or

$$X = 2.09455146.$$

This version of Newton's method appeared first in the 1685 *Algebra* of J. Wallis, and Wallis also used this example. What is described here is substantially different from what we call Newton's method. At each stage it considers a new function, so the method is not iterative. What we call Newton's method, an iterative method, was described by J. Raphson in *Analysis aequationum universalis* of 1690, and Cajori made a case in [Cajori,1911] for calling it the Newton-Raphson method.

Whiteside, [Whiteside,1974, pp. 316–318, Note 151], in agreement with the Adams quotation above, stated that the procedure Newton used on KE in the third edition of the *Principia* is equivalent to the Newton-Raphson method. And he wrote,

> ...Newton had sketched an equivalent application of this 'Newton-Raphson' iterative procedure to the speedy calculation of n-th roots (where $f(x) \equiv x^n - A = 0$) for the benefit of a London table-computer in July 1675...

Raphson restricted his attention to polynomial equations, and in [Cajori,1911, p. 32], Cajori credited Thomas Simpson, [Simpson,1740, p. 81], with the extension of the method to irrational and transcendental functions. But later Cajori in [Newton,1934, Volume II, pp. 647–648, Note 23] acceded to the Adams point of view. So, while many modern textbooks continue to point out that the label "Newton-Raphson" is more accurate for the method, we may rest easy that Newton's method is really *Newton's* method.

Along with Simpson, Euler, Laplace, Lacroix, Gauss, and Legendre all described Newton's method without reference to Newton or Raphson. The first to establish a criterion for convergence of the method was J. Mourraille in *Traité de la resolution des équations en général, Premiere partie:*

Des équations invariables of 1768. J. Fourier also did so in 1818. Cajori, [Cajori,1911, p. 32], wrote that Newton's name became reattached to the method as a result of Fourier's papers. In Whiteside's judgement, [Whiteside,1974, pp. 317–318, Note 151], it was not until [Adams,1882] that Newton's priority was recognized.

Through various editions of the *Principia,* and in other writings correcting and refining some of its details related to KE, Newton actually presented three solutions of KE. The first was Wren's 1658 cycloid solution, presented earlier, and the third was "Newton's method." The second is described in the Scholium to Proposition XXXI; it was a geometric method, hard to follow, which appeared with some errors in the first edition of the *Principia.* Newton made a sequence of amendments and corrections to it, and Whiteside, [Whiteside,1974, pp. 311–314, Notes 135–143], provided a detailed guide to what he called a "clever geometric approximation" and showed its basis in approximation to a parabolic case. D.J. Keill, [Keill,1713], described this method more fully than did Newton and included more examples. However, " ... Newton fell ultimately out of love with this ingenious but impractical geometrical mode of construction, and suppressed it from the later editions of his *Principia. . .*" [Whiteside,1974, p. 314, Note 143]. What superseded it, as Adams described, is the now-familiar Newton's method.

Evidently Keill's attitude changed also, for in his *Introductio ad veram astronomiam* of 1718 he described quite a different method. Since $M - E = e \sin E$ and

$$\sin E = \sin M \cos(M - E) - \cos M \sin(M - E)$$

$$= \sin M \sum_{k=0}^{\infty} \frac{(-1)^k}{(2k)!}(M - E)^{2k}$$

$$- \cos M \sum_{k=0}^{\infty} \frac{(-1)^k}{(2k + 1)!}(M - E)^{2k+1}$$

we may write

$$M - E = e \sin M \sum_{k=0}^{\infty} \frac{(-1)^k}{(2k)!}(M - E)^{2k}$$

$$-e \cos M \sum_{k=0}^{\infty} \frac{(-1)^k}{(2k)!}(M - E)^{2k+1}.$$

Keill wrote

$$\frac{e \sin M}{1 + e \cos M} = (M - E) + \frac{1}{2} \frac{e \sin M}{1 + e \cos M}(M - E)^2$$

$$-\frac{\cot M}{3!}\frac{e\sin M}{1+e\cos M}(M-E)^3$$

$$-\frac{1}{4!}\frac{e\sin M}{1+e\cos M}(M-E)^4\ldots$$

and used series inversion to write

$$E = M - \left(\frac{e\sin M}{1+e\cos M}\right) + \frac{1}{6}\left(\frac{e\sin M}{1+e\cos M}\right)^3\ldots$$

A. Robertson, [Robertson,1816], later made a variation on this method of Keill by writing E as a correction $E_C + z$ to the Cassini approximation, E_C. Following steps much like Keill's, he expressed the correction z as a series in powers of $(e\sin E_C)/(1+e\cos E_C)$:

$$E = E_C + \frac{e\sin E_C}{1+e\cos E_C}$$

$$+ \frac{1}{2!}\left(\frac{e\sin E_C}{1+e\cos E_C}\right)^4$$

$$- \frac{1}{3!}\left(\frac{e\sin E_C}{1+e\cos E_C}\right)^5 + \cdots .$$

Neither result is related to Newton's method, however.

In 1738, in a paper entitled, "The solution of Kepler's problem," [Machin, 1738], J. Machin claimed that all efforts to solve KE other than Kepler's own had been useless or ineffective. Given e, M, and a starting value E_0 for E, he calculated $M_0 = E_0 - e\sin E_0$ and used

$$E_1 = E_0 + \frac{a}{r_0}(M - M_0)$$

as his next approximation of E, where a is the planet's semimajor axis and r_0 is its distance from the sun at the moment its eccentric anomaly is E_0. Since $r_0 = a(1 - e\cos E_0)$, it can be seen that Machin's scheme is precisely the first iteration of Newton's method.

The novelty of Machin's treatment lay in his choice of E_0: let

$$n = \sqrt{5 + \sqrt{16 + \frac{9}{e}}},$$

solve

$$M = n\left[(1-e)s + \frac{e(n^2-1)+1}{3!}s^3\right] \tag{4.7}$$

for s, and take $E_0 = n \arcsin s$. Machin gave no explanation for his choice, and without Brinkley's guidance, [Brinkley,1803, pp. 109–114], we could never have explained it. (We have put the details in Appendix C for those who are curious to see them.)

Machin's E_0 is not convenient to calculate, but it might be thought more useful if e is near 1. Brinkley himself was interested enough to write an algorithm to solve (4.7) with logarithms for such cases. If we view Machin's method, with two centuries' hindsight, as an effort to choose E_0 shrewdly enough that one iteration of Newton's method will suffice to produce an accurate approximation of E which is robust for all choices of e and M, then in fact Machin was by many decades the first player in a game which attracted many participants after 1965. We describe this game in the section "Numerical Experiments with KE" in Chapter 8. In Appendix C we present some calculations we made to measure the accuracy of Machin's E_1 as a solution of KE over a range of values of e.

There are quite a few additional papers which use Newton's method or variations of it in connection with KE, many of which are otherwise not of great interest. The paper of Encke, mentioned earlier in connection with Cassini's approximation, [Encke,1850], presented several computational solutions of KE. The last and most accurate of them is equivalent to two iterations of Newton's method on an initial guess, followed by a linear interpolation.

An interesting approach to solving KE was described by E. Schubert, [Schubert,1854]: how might one choose initial guess E_0 for a solution so that one iteration of Newton's method produces an approximate solution

$$E_1 = E_0 - \frac{(E_0 - e \sin E_0 - M)}{1 - e \cos E_0}$$

of desired accuracy for all M and e?

Schubert's own details were not interesting or conclusive, but his question anticipated by over 100 years a game of numerical experimentation which acquired a number of enthusiastic players. Undoubtedly Machin may be called the first player, and our calculations in Appendix C indicate that his E_0 leads to values E_1 with accuracy on the order of 2×10^{-2}.

As we have noted, there were many mentions and uses of Newton's method made with no mention of Newton. Our candidate for the latest paper to use Newton's method in the service of KE without mentioning Newton is [Sconzo,1933b]. P. Sconzo wrote a series of three papers on KE containing nothing which hadn't been known for a long time. In addition to not mentioning Newton in [Sconzo,1933b], Sconzo scored a rare double by devoting [Sconzo,1934] to the Bessel-Fourier solution (3.8) without a single mention of Bessel.

Ivory's Geometric Iteration

J. Ivory, [Ivory,1805], hoped to find E in KE by a sequence of geometric constructions. His paper was entitled, "A new and universal solution of Kepler's problem,"—the "universal" aspect being his hope that the method would serve as well in cases when e is near 1 and so be useful for comets as well as planets.

Ivory's sequence of constructions bears a superficial similarity to the later process of Encke described in Chapter 2 (which started with the Cassini approximation and refined it by drawing tangents and parallel lines.) Ivory's method, however, used no initial guess for E and drew no tangent lines.

In Figure 4.1 let ADB be the eccentric circle with radius CA equal to the semimajor axis. If M is the mean anomaly and e is the eccentricity, mark F_0 on the ray from C with central angle $M/2$ relative to perihelion A so that

$$|F_0C| = \frac{|CA|}{e}.$$

Draw a line from F_0 to a point H_0 on the diameter through A so that line F_0H_0 intersects vertical ray CD at a point G_0 satisfying $|G_0H_0| = |CA|$.

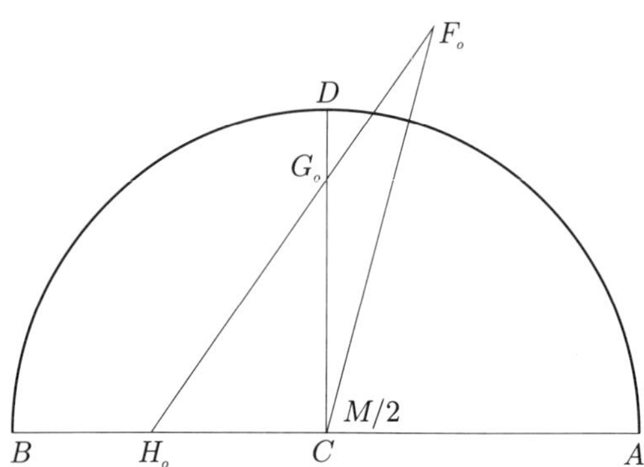

Figure 4.1

Take

$$\angle F_0 H_0 C = \frac{E_1}{2}$$

as a first approximation of $E/2$. Now mark F_1 on the ray CF_0 so that

$$|F_1C| = \frac{\frac{1}{2}(M - E_1)}{\sin\left(\frac{1}{2}(M - E_1)\right)} \frac{|CA|}{e}$$

and in the same manner determine points H_1 on AB and G_1 on CD so that line F_1H_1 intersects CD at G_1 and $|G_1H_1| = |CA|$. Call

$$\angle F_1H_1C = \frac{1}{2}E_2.$$

Henceforth on the ray CF_0 let

$$F_{n+1} = \frac{\frac{1}{2}(M - E_n)}{\sin\left(\frac{1}{2}(M - E_n)\right)}\frac{|CA|}{e}$$

determine H_{n+1} on AB and G_{n+1} on CD so that the line $F_{n+1}G_{n+1}H_{n+1}$ satisfies $|G_{n+1}H_{n+1}| = |CA|$ and set

$$\angle F_{n+1}H_{n+1}C = \frac{1}{2}E_{n+1}.$$

Ivory made a fairly detailed argument to show

$$\{E_{2k+1}\}_{k=0}^{\infty}$$

is increasing and

$$\{E_{2k}\}_{k=0}^{\infty}$$

is decreasing, while

$$\frac{1}{2}|M - E_n| < \rho$$

where ρ is the first quadrant solution of the equation

$$\rho = \sqrt{1 - \frac{e^2}{4}}\tan\rho.$$

(He didn't make any effort to show $|E_{2k+1} - E_k| \to 0$, however.)

If we presume that in Ivory's scheme $E_n \to E$, $F_n \to F$, $H_n \to H$, and $G_n \to G$, then in the limiting case

$$|CF| = \frac{\frac{1}{2}(M - E)}{\sin\left(\frac{1}{2}(M - E)\right)}\frac{|CA|}{e} \tag{4.8}$$

and

$$\angle FHC = \frac{E}{2}$$

where $|GH| = |CA|$.

That this limit E is actually a solution of KE in the form $M = E + e\sin E$ when $e < 1$ may be argued in the following steps:

$$\frac{|CF|\sin\frac{M}{2}}{|CA|\cos\frac{E}{2} + |CF|\cos\frac{M}{2}} = \tan\frac{E}{2}$$

$$|CF| \sin \frac{M}{2} = |CA| \sin \frac{E}{2} + |CF| \tan \frac{E}{2} \cos \frac{M}{2}$$

$$|CF| \left(\sin \frac{M}{2} \cos \frac{E}{2} - \sin \frac{E}{2} \cos \frac{M}{2} \right) = |CA| \sin \frac{E}{2}$$

so

$$|CF| = \frac{|CA| \sin \frac{E}{2} \cos \frac{E}{2}}{\sin \left(\frac{1}{2}(M - E) \right)} = \frac{|CA| \frac{1}{2} \sin E}{\sin \left(\frac{1}{2}(M - E) \right)}. \tag{4.9}$$

From (4.8) and (4.9)

$$\frac{1}{2}(M - E) = \frac{1}{2} e \sin E.$$

Ivory's constructions and arguments required some patching in special cases—for example, when

$$M = \frac{\pi}{2}.$$

With many clever but often tedious details, he treated a variety of cases, including very small e and e very near 1. In several of the cases his details required him to solve a cubic equation and to do his calculations with a parabolic orbit having the same perihelion distance as the elliptic orbit he started with. Ivory was not the first to use such "parabolic" approximations, and we'll say more about them in the next chapter.

Although Ivory claimed considerable accuracy for his method in the examples to which he applied it, his method was far from universal and his results were not very convenient for calculation.

Chapter 5

Solutions of KE for High Eccentricity

Most of the methods of solving KE we have described so far were intended for the cases when e is quite small, as it is for the planets. However, many periodic comets and asteroids have orbits whose eccentricities exceed 0.5, and in these cases the methods we've discussed may be quite unsatisfactory. For example, the famed Halley's Comet has $e = 0.9673$, that of Encke $e = 0.847$, while asteroid Icarus has $e = 0.828$.

Some of the methods we've described are shaky even for the planet Mercury, which of the first six planets has the largest eccentricity, $e = 0.206$. With the triumph of Halley in predicting successfully the return in 1758 of what we now call Halley's Comet, and with increased interest in many other comets and larger asteroids during the 19th century, methods for solving KE when e is large were sought and welcomed.

Barker's Equation and Parabolic Approximations

If an object traces a plane parabolic orbit about the sun as focus, with perihelion distance p, then the polar coordinates (r, v) of the object's position relative to the sun satisfy

$$r = \frac{2p}{1 + \cos v} \tag{5.1}$$

when we take $v = 0$ for $r = p$.

From Newton's laws of motion, of which (5.1) is a consequence, may also be derived another, very special equation which many call *Barker's equation:*

$$\tan \frac{v}{2} + \frac{1}{3} \tan^3 \frac{v}{2} = \frac{t}{2p^{3/2}}. \tag{5.2}$$

The derivation of (5.2) may be classified as elementary but not easy. (A nice reference for the derivation is [Roy,1988, pp. 89–90] but one may need to

57

start at page 70 to see it all.)

T. Barker, after whom this equation is named, was by some 50 years a latecomer to the subject. He wrote a monograph, [Barker,1757], in which parabolas were studied, both as orbits of celestial objects and as paths of projectiles, and in which were included extensive tables. Although the return of Halley's Comet in 1758 made Barker's monograph most timely, it is likely that H. Englefield's *Orbits of Comets* of 1793, which credited Barker's efforts and used his tables, attached Barker's name to equation (5.2).

We believe it was Edmund Halley, [Halley,1705], who first used (5.2) in studying comets and in deducing that the comets of 1531, 1607, and 1682 were the same object. Halley predicted that this object would return in 1758. Its return that year was first observed on Christmas Day by an amateur astronomer, J. Palitzsch, [Moore,1987, pp. 68–69].

Equation (5.2) is also described occasionally as the "parabolic form of Kepler's equation." Although this is suggestive, it can also be somewhat confusing since (5.2) relates time to the true anomaly rather than an eccentric anomaly. There are at least two scenarios which allow one to "see" (5.2) as a legitimate form of KE corresponding to the case $e = 1$. One is to consider the polar form of the equation for position in an elliptical orbit

$$r = \frac{a(1 - e^2)}{1 + e \cos v} \tag{5.3}$$

and to let $a \to \infty$, $e \to 1$ in such a way that $a(1 - e^2) = 2p$ remains fixed. The same dynamical argument which leads from (5.3) to the elliptic form of KE will produce (5.2) from (5.1). Details may be found in [Taff,1985, pp. 44–45]. A second attack is described in [Wintner,1941, p. 194] in an analytical setting for celestial mechanics where the variables are complex and eccentric anomaly is simply a transformation of time; in this setting a form of Barker's equation corresponding to $e = 1$ emerges as one of the three allowable forms of KE. (Further treatment of the various forms KE may take and their integration into "universal" forms of KE is made in Chapter 8 and Appendix I.)

Equation (5.2) may be solved for

$$\tan \frac{v}{2}$$

as a function of t through a pair of auxiliary variables. If

$$\tan \frac{v}{2} = 2 \cot 2w$$

(5.2) is transformed to

$$\frac{1}{3}(\cot^3 w - \tan^3 w) = \frac{t}{2p^{3/2}} \tag{5.4}$$

and, if

$$\cot^3 w = \cot\left(\frac{y}{2}\right)$$

(5.4) is further transformed to

$$\cot y = \frac{3t}{4p^{3/2}}. \tag{5.5}$$

From t one determines $\cot y$, $\cot w$, and $\tan(v/2)$, and then (r, v) may be obtained from tables.

When t measures time since perihelion passage for an elliptical orbit with e near 1 satisfying polar equation (5.3), it is an attractive idea to consider a parabolic path of perihelion $p = a(1 - e)$ matching that of the ellipse and to solve (5.2) for a value $v = u$ corresponding to t.

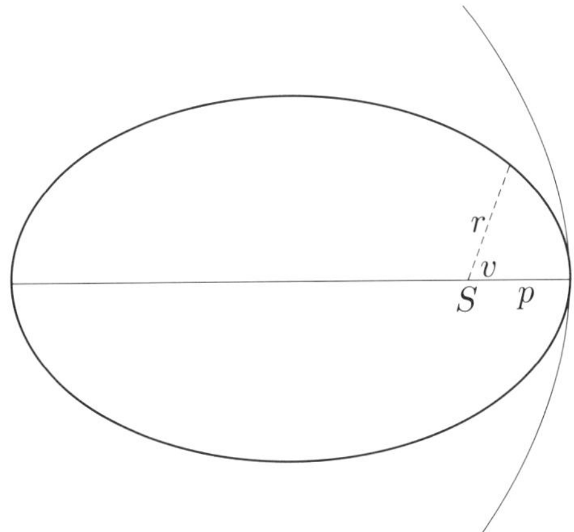

Figure 5.1

Then

$$\left(\frac{a(1 - e^2)}{1 + e\cos u}, u\right)$$

may serve as an approximation to position in the elliptic orbit. For other than small values of t, the approximation of v by u calls for some adjustment.

Perhaps the first to describe this method and a means of correcting it was T. Simpson, [Simpson,1740]. A complete account of what Simpson did is found in an Appendix written by N. Bowditch to appear after Book VII of his translation and commentary to Laplace's *Mécanique céleste*, [Laplace,1966, Volume III, pp. 761–781]; some appreciation of the errors of Simpson's work is made in [Brinkley,1803].

Simpson used (5.2) to determine an initial value of $v = u$. To adjust u he let the true anomaly be $v = u + x$ and determined x in the following manner. From the relation (1.6)

$$\tan\frac{v}{2} = \sqrt{\frac{1+e}{1-e}}\,\tan\frac{E}{2}$$

he expressed E as a series in powers of $\tan(v/2)$. Since, through KE, t is a function of E, he expressed t in powers of $\tan(v/2)$, neglecting powers of $(1-e)$ higher than the first, as

$$t = (\text{constant})\tan\frac{v}{2}\left[1 + \frac{1}{3}\tan^2\frac{v}{2} + (1-e)\left\{\frac{1}{4} - \frac{1}{4}\tan^2\frac{v}{2} - \cdots\right\}\right]. \quad (5.6)$$

With $v = u + x$, and ignoring all but the first-degree terms in x, he wrote

$$\tan\frac{v}{2} + \frac{1}{3}\tan^3\frac{v}{2} = \tan\frac{u}{2} + \frac{1}{3}\tan^3\frac{u}{2} + \frac{1}{2}x\sec^4\frac{u}{2}$$

and took

$$\left\{ \begin{array}{c} x = \left(\dfrac{1-e}{10}\right)\tan\dfrac{u}{2}\left(4 - 3\cos\dfrac{u}{2} - 6\cos^4\dfrac{u}{2}\right) \\[4mm] v = u + x. \end{array} \right\} \quad (5.7)$$

Simpson's approximation of v was satisfactory for e very close to 1. But for Halley's Comet, where $1 - e$ is about 0.03, the size of the terms involving $(1-e)^2$ which Simpson omitted can be significant, and Simpson's method had considerable error.

Ivory, [Ivory,1805], also treated KE for e near 1 by means of a parabolic path and equation (5.2). In the same notation used above, Ivory's approximation of v was

$$v = u + \frac{1}{15}(7 - 6\cos u - \cos 2u)\left(\frac{1-e}{1+e}\right)\tan\frac{u}{2}$$

$$+\frac{1}{3150}(510 - 78\cos u - 341\cos 2u$$

$$-84\cos 3u - 7\cos 4u)\left(\frac{1-e}{1+e}\right)^2\tan^3\frac{u}{2}.$$

The improvement of Simpson's method specifically for Halley's Comet was the subject of Bessel's first paper in astronomy, [Moore,1987, p. 18]. In [Bessel,1805] he undertook to add to Simpson's approximation $v = u + x$ the contribution of terms with factor $(1-e)^2$ which Simpson had omitted. The resulting approximation by Bessel is easier to follow from Bowditch's details, [Laplace,1966, Volume III, pp. 774–775]. Bessel took

$$v = u + x + (1 - e)^2 B \qquad (5.8)$$

where, with

$$\theta = \tan \frac{u}{2},$$

$$B = \frac{\left[-\frac{1}{16}\theta - \frac{9}{16}\theta^3 + \frac{37}{80}\theta^5 + \frac{531}{560}\theta^7 + \frac{13}{35}\theta^9 + \frac{9}{350}\theta^{11}\right]}{(1 + \theta^2)^4}.$$

For Halley's Comet (5.8) is more satisfactory than (5.7), but when $1 - e$ exceeds 0.03 its accuracy falters.

Gauss' Method

For the large gap $0.2 < e < 0.95$ neither the methods of the earlier chapters nor those detailed above are satisfactory. The great C.F. Gauss conceived a method of iteration based on a parabolic approximation which is effective for values of e in this gap. The method is not simple to derive or to explain. Although the method is clever and robust in e, its use requires auxiliary tables. As we'll see after we describe the method, Gauss' method was much-admired but not much-used.

Gauss' method is one to determine v directly in terms of time t since perihelion passage for an elliptical orbit.

From v, of course, one may determine E, but it is only in this indirect sense that Gauss' method solves KE.

If the elliptical orbit has perihelion distance p, Gauss considered a fictitious parabolic orbit with perihelion distance

$$p_1 = p \left[\frac{B^2}{1 - .9(1 - e)}\right]^{1/3} \qquad (5.9)$$

where B is a number slightly larger than 1 to be described later. It was intended that one have available several auxiliary tables to look up parameters in the method as needed. One of these auxiliary tables was a table of solutions of (5.2) in terms of the "perihelion" p_1. Our description of the parameters of the method is taken from Bowditch's Appendix in [Laplace,1966, Volume III].

Presume, for suitable constant k, that KE is written

$$kt \left(\frac{1 - e}{p}\right)^{3/2} = E - e \sin E. \qquad (5.10)$$

Let

$$\alpha = 1 - e, \quad \alpha' = \sqrt{.1 + .9e}$$
$$\beta = \frac{\alpha}{2\alpha'^2}$$
$$\gamma = \sqrt{\frac{5(1+e)}{1+9e}}$$

and define

$$\left.\begin{array}{c} T = \tan^2 \dfrac{E}{2} = \dfrac{1-e}{1+e} \tan^2 \dfrac{v}{2} \\[2ex] A = \dfrac{15(E - \sin E)}{9E + \sin E} \\[2ex] B = \dfrac{9E + \sin E}{20\sqrt{A}} \\[2ex] C = \dfrac{A}{T} + \dfrac{4}{5}A. \end{array}\right\} \tag{5.11}$$

By means of series expansions and inversions of series it will follow that

$$\left.\begin{array}{c} A = \dfrac{1}{4}E^2 - \dfrac{1}{120}E^4 - \dfrac{1}{20160}E^6 - \cdots \\[2ex] \sqrt{A} = \dfrac{1}{2}E - \dfrac{1}{120}E^3 - \dfrac{1}{84000}E^5 - \cdots \\[2ex] B = 1 + \dfrac{3}{2800}E^4 - \cdots \\[2ex] A = T - \dfrac{4}{5}T^2 + \dfrac{24}{25}T^3 - \dfrac{1592}{2625}T^4 + \cdots \\[2ex] \dfrac{A}{T} = 1 - \dfrac{4}{5}A + \dfrac{8}{175}A^2 + \dfrac{8}{525}A^3 + \dfrac{1896}{336875}A^4 + \cdots \\[2ex] C = 1 + \dfrac{8}{175}A^2 + \dfrac{8}{525}A^3 + \dfrac{1896}{336875}A^4 + \cdots \end{array}\right\} \tag{5.12}$$

The quantities E, A, B are related by

$$9E + \sin E = 20\sqrt{A}B$$

and in (5.10)

$$\left\{ \begin{aligned} &kt\left(\frac{1-e}{p}\right)^{3/2} = E - e\sin E \\[2ex] &\qquad = (9E + \sin E)\left\{\frac{1-e}{10} + \right. \\[2ex] &\qquad\qquad \left. \left(\frac{1+9e}{10}\right)\left[\frac{E-\sin E}{9E+\sin E}\right]\right\} \\[2ex] &\qquad = 2B\left[(1-e)\sqrt{A} + \left(\frac{1+9e}{15}\right)A^{3/2}\right]. \end{aligned} \right\} \qquad (5.13)$$

If we let

$$\sqrt{A} = \sqrt{\frac{5(1-e)}{1+9e}}\,\tan\frac{w}{2}$$

or

$$A = \beta\tan^2\frac{w}{2}$$

equation (5.13) is transformed into a copy of Barker's equation (5.2):

$$\frac{kt}{p^{3/2}} = 2B\left(\frac{5}{1+9e}\right)^{1/2}\left(\tan\frac{w}{2} + \frac{1}{3}\tan^3\frac{w}{2}\right)$$

or

$$\frac{\alpha' t}{Bp^{3/2}} = \frac{\sqrt{2}}{k}\left(\tan\frac{w}{2} + \frac{1}{3}\tan^3\frac{w}{2}\right).$$

From (5.9)

$$\frac{\alpha'}{Bp^{3/2}} = \frac{(.1+.9e)^{1/2}}{Bp^{3/2}} = \frac{1}{p_1^{3/2}}$$

so (5.10) is rewritten as

$$\frac{t}{p_1^{3/2}} = \frac{\sqrt{2}}{k}\left(\tan\frac{w}{2} + \frac{1}{3}\tan^3\frac{w}{2}\right)$$

which is a copy of (5.2) with "fictitious" p_1.

To make the method work, two auxiliary tables are required. The first is a table of values of A, B, C, T, which may be compiled from (5.11) in terms of values of E between 0 and π chosen as densely as one wishes to pay for in computations. The second is a table of solutions of (5.2) for a variety of values

$$\frac{t}{2p^{3/2}}.$$

For reference we call these Table 1 and Table 2, respectively.

As we see in (5.12), the largest value of B when $0 \leq E \leq \pi$ will be no greater than 1.105. Now we can describe an algorithm for Gauss' method.

Suppose e, k, t are given. To find v:

(a) Start with $B = 1$ and calculate p_1.

(b) Use Table 2 to determine w and then calculate

$$A = \beta \tan^2 \frac{w}{2}$$

$$\frac{A}{T} = 1 - \frac{4}{5}A + \frac{8}{175}A^2 + \cdots$$

$$C = \frac{A}{T} + \frac{4}{5}A$$

$$T = \frac{A}{C - \frac{4}{5}A}$$

(c) Use Table 1 to obtain an improved value of B and p_1.

(d) Repeat (b) with new B, p_1.

(e) Determine v from

$$\tan \frac{v}{2} = \sqrt{\frac{1+e}{1-e}}\sqrt{T}$$

$$= \frac{\gamma \tan \frac{w}{2}}{\sqrt{C - \frac{4}{5}A}}$$

and then obtain E from

$$\tan \frac{E}{2} = \sqrt{T}.$$

The loop $[(b) \Rightarrow (c) \Rightarrow (d)]$ may be repeated more than once; except for E near 0 or π it was expected that one passage through steps (a) – (e) would suffice to find v and E with sufficient accuracy. Were observations to allow a better starting value for B than 1, the sequence $[(a) \Rightarrow (b) \Rightarrow (e)]$ might suffice.

Gauss presented this intricate method in 1809 in his influential *Theoria motus,* [Gauss,1809, Sections 35–43]. In our own examination of the literature of KE for the period 1810–1940 we find the method of Gauss mentioned frequently but seldom used. As we'll detail in Chapter 7, there was no lack of activity toward approximate solutions of KE which might be used quickly and repeatedly, many of which involved the use of auxiliary tables. The examples

Gauss gave for his method were used often as test cases for comparison with new methods either on grounds of accuracy or of convenience or both. Implicitly the results of Gauss' own calculations in these examples were made a standard for accurate solutions which other methods would try to meet.

A typical example may be found in [Brunnow,1858]. After acknowledging the efficacy of Gauss' method when e is not small, F. Brunnow expressed the hope of finding a more convenient method. What he had in mind was to improve on Bessel's refinement of Simpson's approximation.

For an elliptical orbit of eccentricity e and perihelion distance p, Brunnow let

$$\lambda = \frac{1-e}{1+e}$$

and

$$u = \tan \frac{v}{2}$$

where

$$\frac{kt}{2p^{3/2}} = \tan \frac{v}{2} + \frac{1}{3} \tan^3 \frac{v}{2} = u + \frac{1}{3} u^3$$

for a parabolic orbit with the same perihelion distance. He wrote

$$\frac{kt}{2p^{3/2}} \sqrt{1+e} = \int \frac{1+u^2}{(1+\lambda u^2)^2} du = U + \frac{1}{3} U^3 \tag{5.14}$$

with a new variable U.

With $U = u + a_1 \lambda + a_2 \lambda^2 + a_3 \lambda^3 \ldots$, Brunnow determined a_1, a_2, a_3 in terms of powers of u. From e and t standard tables would supply U in (5.14). For $V = 2 \arctan U$ and series calculations incorporating all terms through λ^3, Brunnow found

$$v = V + \left[\frac{\frac{4}{3} U^3 + \frac{4}{5} U^5}{(1+U^2)^2} \right] \lambda$$

$$+ \left[\frac{\frac{22}{15} U^5 + \frac{598}{315} U^7 + \frac{86}{105} U^9 + \frac{18}{175} U^{11}}{(1+U^2)^4} \right] \lambda^2$$

$$+ \left[\frac{\frac{584}{315} U^7 + \frac{9752}{2835} U^9 + \cdots + \frac{184}{7875} U^{17}}{(1+U^2)^6} \right] \lambda^3 .$$

With this "convenient" result, Brunnow obtained v in the same example Gauss had used for Halley's Comet to within $0.02''$ of the value Gauss had gotten. (Simpson's method in this instance had produced a value differing from Gauss' by more than $5''$.)

Through additional elaborate changes of variables and specialized auxiliary tables, both Th.v. Oppolzer, [Oppolzer,1879,1880], and G.M. Searle, [Searle,1889,1892], sought to express $\tan(v/2)$ in terms of e, t, and Brunnow's λ in hopes of obtaining accurate values for high-eccentricity orbits. It is difficult to find any evidence these efforts were actually applied to any calculations except those examples actually used by Gauss. Indeed the very choices of intermediate parameters and terms to be neglected as Searle made them seemed designed only to produce high agreement with the Gauss examples.

High eccentricity cases were considered further between 1900 and 1940, but Gauss' method found little favor until mechanical calculators, and later computers, became available. In a privately-published 1948 memoir entitled *"The Computation of Orbits,"* P. Herget explicated Gauss' method and refined it for machine calculation. This material was reworked in [Benima,1969] to provide a modern, computer-implementable version of Gauss' method.

Chapter 6

Cauchy and KE

Were we, in the style of the physicist, to label all the items in our mathematical catalogue with proper names, we would require a large supply of labels marked "Newton," "Euler," "Gauss," and "Cauchy." In the context of KE and its neighborhood of celestial mechanics we have already encountered the first three names: Newton's prominently, Euler's peripherally, and Gauss' technically. This chapter is about Augustin Cauchy, but more than what he did to KE or about KE it is an account of how what he did influenced and altered what others did to KE.

As much for his specific theorems and results in mathematics, Cauchy is renowned for the efforts he made to give to analysis methods and foundations adequate for rigor and proof. The age of rigor in analysis is generally acknowledged to have begun around 1820 with Cauchy as its central figure. He was neither the first nor the only one to recognize the need for making precise the basic concepts and for establishing standards to ensure the validity of results obtained by formal means. Nor were all the ideas to which his name has become attached original with him. Yet, to a great extent, the fundamental concepts, processes, and "rules of the road" for elementary calculus, classical analysis, and modern analysis today trace directly back to him and are practised as he intended. Of the many references to which one might turn to add to this very brief appreciation of Cauchy, we'll cite a short one, [Boyer,1989, pp. 572–580], a fuller one, [Grabiner,1981], and one containing some original sources, [Birkhoff,1973].

In addition, Cauchy enlarged both the scope and the machinery of analysis from real-valued functions of a real variable to complex-valued functions of a complex variable. The gains in richness, clarity, and technique which result when real function theory is embedded in complex function theory far exceed the costs involved in intuition, visualization, and increased abstraction. These gains, together with the vast body of complex-variable techniques in applied mathematics, and the results in real analysis whose only known proofs require complex analysis, are part of Cauchy's legacy.

We will detail how celestial mechanics, and in particular KE and Lagrange's theorem, were on the scene, if not at the heart, of Cauchy's development of both a theory of convergence for power series and a theory of analytic functions of a complex variable. The suggestion that KE played a substantial role in Cauchy's work has appeared in several references, namely, [Taff,1985, p. 82], [Wintner,1941, p. 217, Footnote], and [Brill,1894, p. 178].

The successes of 18th century mathematics created many areas in which the means to resolve contradictory results and to establish validity were not yet available. Of these areas, the treatment of infinite series was most prominent, and it is in this area that Cauchy's work and KE meet. The point of intersection is Lagrange's theorem and the resulting solution of KE as a power series in powers of e. Neither in his 1771 paper on KE nor in his 1770 paper which stated the theorem had Lagrange made any effort to assess the validity or convergence of the infinite series he obtained in a purely formal manner. In the earlier volumes of his *Mécanique céleste* Laplace had used Lagrange's theorem to obtain power series solutions for E and powers of r. Laplace attempted arguments to obtain the region of convergence for his power series; while the results were generally correct, the arguments were not secure.

Formal operations on power series were commonly used and not subjected to much critical analysis, the *de facto* standard being whether useful conclusions were obtained. It became Cauchy's thesis that arguments based on infinite series which made no effort to establish their convergence were not to be trusted. Of the many areas of mathematics and applications in which infinite series were used, celestial mechanics was only one. Whether as a matter of convenience or one of shrewd choice, Cauchy used celestial mechanics, Lagrange's theorem and KE as a setting in which to develop his ideas and theorems for establishing a theory of convergence of power series. Perhaps the importance and currency of Laplace's *Mécanique céleste* drew Cauchy's attention to that setting. In both [Kline,1972, p. 972] and [Boyer,1989, p. 578] there appears a story that Cauchy's reading of his first paper on convergence of infinite series sent Laplace home to check that all the series he used in the *Mécanique céleste* were actually convergent in Cauchy's sense. (They were!)

To learn what Cauchy did, when he did it, and something of his motivations requires time and the study of experts. Cauchy was moved to write early and often about his ideas and progress. His collected works comprise 26 volumes which are incomplete and whose arrangement defies explanation, but a two-volume biography of Cauchy helps somewhat with the chronology, [Valson,1868]. Many of Cauchy's papers were published long after their initial versions had been written and perhaps read at meetings, and the published versions often had undergone revisions, additions, and corrections which make it difficult to assign accurate dates to specific results. Also, in response to political and religious affairs during a very turbulent era, Cauchy exiled himself from France for more than one extended period, and his writings during these times are not completely known or available. The memoir [Cauchy,1831], for example, which is a key paper for our story, was written while Cauchy was in Turin; the version which is available is a revision in [Cauchy,1840a, Volume 2, pp. 41–112]. Our account is based on review articles by A. Brill and M. Noether, [Brill,1894, Chapter II], W.F. Osgood, [Osgood,1905], and Wintner's book, [Wintner,1941]. Understating the profusion of Cauchy's papers on celestial mechanics, we cite only a small number which seem to bear most directly on KE.

Cauchy's Treatment of Lagrange's Theorem and KE

In Chapter 3 we described Lagrange's theorem and the solution (3.2) of KE consequent to it. Laplace in [Laplace,1799] did make an argument to conclude (3.2) converges when $e < 0.66195$ (the correct value being $0.66274\ldots$). In [Laplace,1823] he used a similar argument for a series to express r in powers of e. In this argument Laplace set

$$M = \frac{\pi}{2}$$

and estimated how large e would need to be for the series to diverge.

Cauchy in [Cauchy,1829a, p. 98] wrote about these results:

> Struck by a result so worthy of notice, I asked myself if it would not be possible to determine generally the conditions of convergence of the Lagrange series.... My studies lead me to recognize that conditions can always be deduced for the solution of a transcendental equation which contains as a particular case the equation considered by M. Laplace.

Here is what Cauchy established. If

$$f(w) = f(a) + \sum_{n=1}^{\infty} \frac{t^n}{n!} \frac{d^{n-1}}{da^{n-1}} \left\{ f'(a)[F(a)]^n \right\} \tag{6.1}$$

where $w = a + t\, F(w)$, let $R(a)$ be the absolute maximum of the modulus of

$$\frac{F(a+x)}{x}$$

over all solutions (including complex solutions) of the equation

$$\frac{d}{dx}\left[\frac{F(a+x)}{x}\right]=0.$$

When

$$S_n=\frac{1}{(n-1)!}\frac{d^{n-1}}{da^{n-1}}\{f'(a)[F(a)]^n\}$$

for large n, there is a constant K, independent of n, for which

$$|S_n|<\frac{K[R(a)]^n}{n^{1/2}}. \tag{6.2}$$

The series

$$\sum_{n=1}^{\infty}S_n t^n$$

and the series

$$a+\sum_{n=1}^{\infty}\frac{1}{n}S_n t^n$$

(which is (6.1)) converge if $R(a)<1$ and diverge if $R(a)>1$.

His method to attain these results, in particular the estimate (6.2), was to express S_n in the integral form

$$\left\{\begin{aligned}S_n&=\frac{1}{2\pi}\int_{-\pi}^{\pi}\frac{f'(a+r\exp(is))\,[F(a+r\exp(is))]^n}{r^{n-1}\exp[i(n-1)s]}\,ds\\[2ex]&=\frac{1}{2\pi}\int_{-\pi}^{\pi}f'(a+r\exp(is))\left[\frac{F(a+r\exp(is))}{(r\exp(is))^{1-\frac{1}{n}}}\right]^n ds\end{aligned}\right\} \tag{6.3}$$

which is, for $z=r\exp(is)$,

$$S_n=\frac{1}{2\pi i}\oint_{\{|z|=r\}}\frac{f'(a+z)[F(a+z)]^n}{z^n}\,dz.$$

In using (6.3) to obtain (6.2), Cauchy replaced the exponent

$$1-\frac{1}{n}$$

by 1 since he intended n to be large.

When we specialize to KE by taking $w = E$, $a = M$, $t = e$, and $F(w) = -\sin E$, convergence of the two series is ensured if the absolute maximum, $R(M)$, of

$$\left| \frac{\sin(M + x)}{x} \right|$$

over all solutions of

$$\frac{d}{dx}\left[\frac{\sin(M + x)}{x}\right] = 0 \tag{6.4}$$

is less than 1.

Now we'll follow the steps Cauchy took to decide that $R(M) < 1$ as long as $e < 0.6627434193$.

If x satisfies (6.4), $\tan(M + x) = x$. Here Cauchy set

$$M = \frac{\pi}{2},$$

so $-\cot x = x$, or

$$\tan x = -\frac{1}{x}. \tag{6.5}$$

With $x = r\exp(is)$,

$$|g(x)| = \left| \frac{\sin(\frac{\pi}{2} + x)}{x} \right|$$

becomes

$$|g(r\exp(is))| = \left(\frac{e}{2r}\right)[\exp(2r\sin s) + \exp(-2r\sin s) + 2\cos(2r\cos s)]^{1/2}$$

which has maximum value

$$\frac{e}{2r}[\exp(r) + \exp(-r)]$$

when

$$s = \frac{\pi}{2}.$$

Then (6.5) becomes

$$\frac{\exp(r) - \exp(-r)}{\exp(r) + \exp(-r)} = \frac{1}{r}. \tag{6.6}$$

Cauchy concluded that (3.2) converges when

$$R\left(\frac{\pi}{2}\right) = \frac{e}{2r}[\exp(r) + \exp(-r)] < 1$$

or when

$$e < \frac{2r}{\exp(r) + \exp(-r)}.$$

Since (6.6) implies

$$\left\{ \begin{aligned} \frac{r}{\exp(r) + \exp(-r)} &= \frac{1}{\exp(r) - \exp(-r)} \\ &= \frac{\sqrt{r^2 - 1}}{2} \end{aligned} \right\} \tag{6.7}$$

it was Cauchy's conclusion that (3.2) converges when $e < \sqrt{r^2 - 1}$ for the solution r of (6.6). Cauchy made no explanation for his choice

$$M = \frac{\pi}{2}$$

in estimating $R(M)$. Both V. Puiseux and J.-A. Serret later undertook to close this gap in Cauchy's argument.

Cauchy took his argument this far in [Cauchy,1829a]. In the remainder of this paper and in [Cauchy,1829b] he made no attempt to calculate r or $\sqrt{r^2 - 1}$. Instead he used his general convergence theorem for series of the form (6.1) to study the binomial theorem and what we now call the generating function for the Legendre polynomials.

It is in [Cauchy,1831] that Cauchy used exponential and binomial series to approximate a solution of equation (6.6) by some trial and error as $r = 1.199678\ldots$ so that $\sqrt{r^2 - 1} = 0.6627434\ldots$.

In this paper Cauchy took a somewhat sterner attitude toward Laplace's failure to justify Lagrange's theorem and to establish more carefully the region of convergence of the series he used in his *Mécanique céleste*. Cauchy's paper is entitled, *Mémoire sur la mécanique céleste et sur un nouveau calcul appelé Calcul des Limites*. In it Cauchy described residue calculations, established how to determine the region of convergence of a power series, obtained estimates for the error when one approximates their sums by a finite number of terms, and gave his argument proving Lagrange's theorem.

Cauchy later wrote many papers dealing with the problems of celestial mechanics which Laplace had treated. Important more for their mathematics than their contributions to celestial mechanics, most of them rest almost completely on the works [Cauchy,1829a,1831].

In [Cauchy,1841], Cauchy treated KE directly, seeking to write

$$E = M + \sum{}^{*} A_n \exp(inM) \tag{6.8}$$

where we use

$$\sum{}^{*}$$

to denote

$$\sum_{\substack{n = -\infty \\ n \neq 0}}^{\infty}$$

Much the way Bessel calculated his Fourier sine coefficients, Cauchy found

$$A_n = \left(\frac{e}{2\pi}\right) \int_0^{2\pi} \{\sin nM \exp(-inM)(1 - e\cos M)\exp(ine\sin M)\}\, dM \quad (6.9)$$

and used residue calculations to assign values to A_n which are equivalent to

$$A_n = \frac{-iJ_n(ne)}{n}, n = \pm 1, \pm 2, \pm 3, \ldots .$$

Letting

$$g(z) = \frac{e}{2}\left(z + \frac{1}{z}\right)$$

and

$$C = \{|z| = 1\}$$

we may rewrite the integrals (6.9) in the form

$$A_n = \left(\frac{e}{2i}\right)\left(\frac{1}{2\pi i}\right)\oint_C \frac{\left(z - \frac{1}{z}\right)[1 - g(z)]\exp[ng(z)]}{z^{n+1}}\, dz. \quad (6.10)$$

These expressions and the Cauchy numbers (see Chapter 3) which are defined in this paper were the starting point for the generalizations of Bessel functions made by Bourget.

By the 1850's Cauchy's papers on celestial mechanics had a more retrospective, expository air. By this time both his calculus of residues and Bessel's functions and their identities were being assimilated and becoming standard techniques in applied mathematics. In [Cauchy,1854a,b] Cauchy returned to the expressions (6.9); the estimates of A_n for e near 1 they allowed were more useful in expressing the solution of KE than those of Carlini for

$$\frac{J_n(ne)}{n}.$$

In [Cauchy,1854b] he treated the special case of KE with $e = 1$, calculating A_1, A_2, A_3, A_4 explicitly and describing several problems whose solutions rest on this special case.

Following Cauchy from 1849 to 1941

Even in the narrow context of celestial mechanics centering on KE there is considerably more to tell about the mathematics Cauchy set in motion.

In arguing that the region of convergence of series (3.2) and others used by Laplace is $e < 0.6627434\ldots$, Cauchy considered only

$$M = \frac{\pi}{2}.$$

V. Puiseux, [Puiseux,1849], considered Cauchy's argument incomplete and undertook to supply a fuller one.

Among all solutions (e, E, M) of the system

$$\left\{ \begin{array}{l} E - e \sin E = M \\ \\ 1 - e \cos E = 0 \end{array} \right\} \tag{6.11}$$

where M is real but e and E are complex, Puiseux sought to calculate $h = \min|e|$ and to argue that (3.2) converges when $e < h$ for all values of M.

If $h(M) = \{\min|e| : (e, E, M) \text{ satisfies } (6.11)\}$, Puiseux proved that

$$h = h\left(\frac{\pi}{2}\right)$$

—that is, the only case Cauchy took was the right one—and $h = 0.6627434\ldots$. (In fact, throughout his paper Puiseux used the incorrect value 0.66195 which Lagrange had given.) He argued that if the minimum $h(M)$ is attained when $e = a + ib$ and $E = A + iB$, then

$$M - A = X\left(M - \frac{\pi}{2}\right)$$

where X is always positive, and

$$\frac{dh}{dM} = \frac{hX\left(M - \frac{\pi}{2}\right)}{(M - A)^2 + B^2}$$

By what calculus students know as the "first derivative test" Puiseux concluded that $h(M)$ has a minimum when

$$M = \frac{\pi}{2}.$$

As Puiseux noted in a later description of the same work, [Puiseaux,1884], series (3.2) may converge for values of $e > 0.6627434\ldots$ when

$$M \neq \frac{\pi}{2}$$

and even for e near 1 if M is near 0 or π. J.-A. Serret also treated the problem of convergence of series (3.2), first in a note, [Serret,1856], which attracted Cauchy's attention, and later more completely in the paper [Serret,1859]. Serret felt he could establish Cauchy's results more simply by studying more carefully the solutions of the system (6.11). Although nominally in the service of KE, Serret's work was more general than Puiseux's in giving attention to the distribution of solutions of (6.11) over the complex e-plane. After Serret's presentation of the 1856 note, Cauchy responded with a paper, [Cauchy,1857], which explained how the logarithmic derivative, f'/f, locates the zeros and poles of a transcendental function f. In the second part of the paper Cauchy applied his results to examine the distribution of complex solutions and multiple solutions of KE and obtained Serret's results by means of the logarithmic derivative.

Both Puiseux and Serret were accomplished mathematicians. Serret's name endures in the Serret-Frenet formulas of differential geometry. Puiseux was influential in the development of complex analysis as a mathematical theory; he was among the first to recognize a distinction between isolated and nonisolated singular points and to consider essential singular points, [Bottazzini,1986, pp. 162–173].

P.L. Tchebycheff, [Tchebycheff,1857], was also attracted to Cauchy's work and the problem of the region of convergence of series (3.2). With considerations not so different from those of Puiseux and Serret, Tchebycheff obtained the number 0.6627434... and the additional information that the error in approximating $E - M$ with N terms of series (3.2) is less than $(0.662743...)^N$.

It is a staple of Cauchy's theory of residues that when f is an analytic function of a complex variable, the number of solutions of the equation $f(z) = 0$ inside a closed curve C on which f does not vanish is counted by the line integral

$$\frac{1}{2\pi i} \oint_C \frac{f'(z)}{f(z)} dz. \tag{6.12}$$

Since the quantity (6.12) is an integer, it is reasonable to expect that if we change f by a small amount on C, we won't change the number of solutions of $f(z) = 0$ inside C. It was E. Rouché, [Rouché,1862], who showed that, when $|g(z)| < |f(z)|$ at all points of C, $f(z) + g(z) = 0$ and $f(z) = 0$ have the same number of solutions inside C. Rouché's theorem is one of the minor gems of a standard introduction to complex analysis. The paper in which Rouché established the theorem is entitled, *Mémoire sur la série de Lagrange*. Here Rouché gave a careful treatment of Cauchy's results for the region of convergence of the series (6.1), applied them to (3.2), obtained the value 0.6627434... solving (6.7) by Newton's method, and extracted his now-famous theorem as a generalization of what he learned from his examination of (3.2) and (6.1). This direct connection of Rouché's theorem to Lagrange's theorem and KE is little-mentioned in modern textbooks, and mentions of it may be found in [Wintner,1941, p. 217, Footnote] and [Osgood,1905, p. 46].

Of later references to (3.2) and the work of Cauchy, Puiseux, and Serret, most add little to what we have seen. An interesting exception is [Moulton,1903]. Given e, suppose E_0, M_0 satisfy

$$M_0 = E_0 - e \sin E_0.$$

Moulton treated a series for $E - E_0$ in powers of $M - M_0$ in the same spirit as Cauchy, Puiseux, and Serret and found its radius of convergence to be

$$\sqrt{M_0^2 + \left[\cosh^{-1}\left(\frac{1}{e}\right) - \sqrt{1 - e^2}\right]^2}.$$

After the 1840's classical analysis in all its meanings developed most vigorously for almost 100 years. One can gain some appreciation of what this meant for celestial mechanics by studying how A. Wintner, [Wintner,1941], chose to display the subject with all the power and sophistication of analysis. In the specific matter of KE and Lagrange's theorem, a complete argument, correct in a modern sense of mathematical rigor, was not made until 1906. (Interestingly, it was not until about this time also that a proof of the Cauchy integral theorem could be made without assuming that analytic functions have continuous first derivatives.)

For each real value of M, solving KE is equivalent to finding an inverse of the function

$$e(E) = \frac{E - M}{\sin E}. \tag{6.13}$$

For this, or any nonrational function of E whose only finite singularities are poles, E. Picard showed that there will be infinitely many solutions E. More than just the points where $1 - e \cos E = 0$ studied by Cauchy need be examined in establishing the region of convergence of series (6.1). It was A. Hurwitz, [Hurwitz,1906], who showed one must consider as well the possible asymptotic values of (6.13)—that is, the possible limits of $e(E)$ as $|E|$ is allowed to grow to ∞ along any continuous curve in the complex E-plane. The details and Hurwitz' results are described in [Wintner,1941, pp. 217–218]. Hurwitz' paper used KE only as a first example to illustrate a result of some generality.

It is probably fair to say that the role of KE in the works this chapter has described was that of a problem, with some claim to importance and independent interest, which provided a worthwhile setting for serious treatment of a collection of related mathematics problems. And to a large extent, the workers involved were scientists with primary allegiance to mathematics. That their results might enhance progress in celestial mechanics would have been a lesser goal than that they might contribute to progress in mathematics.

Over roughly the same period, 1820–1920, there was emerging a rather separate group, for whom solving KE was a recurring chore in the accomplishment of which rapidity and convenience were prized, but elegance of process was not. The collection of approximations, auxiliary tables, and analogue devices this group came to develop to lighten the computational load will be described in the next chapter.

Chapter 7

Calculations, Auxiliary Tables, and Analogue Devices

In contrast to work described in the previous chapter, there was a large collection of work after 1855 devoted to solving KE which made little use of, and contributed little to, the elegant mathematics emerging from the study of the Kepler problem. This work hoped to ease the practical problems of establishing the positions of astronomical objects and determining their orbits. These problems involved fierce and repetitive calculations and needed to be tamed by some kind of convenience and accuracy so that celestial mechanics might make progress in understanding the data of more and better observations.

Behind the problems associated with orbit determination there were several broad areas which motivated portions of the work of celestial mechanics in the 19th century. These were the study of asteroids, or "minor planets," commencing with the discoveries of Ceres and Pallas in 1801 and 1802, the growing studies of binary stars, and the continuing discourse on the classification and origins of comets. From roughly 1800 to 1870 there were few who questioned Laplace's thesis that comets must be of interstellar origin, and those few attracted little support. As more comets were observed after 1850, including more periodic comets of short period, there was more reason to argue there could be several regions of origin of comets, including some in the planetary system. The recent book by M.E. Bailey et al., *The Origin of Comets*, [Bailey,1990, pp. 124–137], gives full attention to this issue.

This chapter gathers some of the work done on KE in the form of approximations, tables, and physical devices proposed to obtain solutions of KE quickly and easily. In reading a great deal of this work, one can't help noticing how different it was from work in mathematics of the time, from earlier work already discussed, and also from present scientific writing. Authors we survey here seemed to know or to have read little of the work already done. They were quick to publish, to make extravagant claims for very slight results, and to claim as new results many which were not. (In the sections "Solution by Cycloid" and "Solution by Curve of Sines" of Chapter 2 we have already described some examples.) It was also common for an author to publish a new result nearly simultaneously in several different journals, countries, and languages. Such "publication padding" is not rare today, although it is not admired, and perhaps we can view it more permissively for a time when journals were not as widely-circulated and accessible as is the case today.

One of the critical side-effects of these shortcomings is that authors were less generous in citing the works of others and in leaving a trail of references than is the modern practise—and was more the practise of contemporary mathematics literature. To describe the work of this chapter it is appropriate to quote L.G. Taff, [Taff,1985, p. 54]: "There is more literature on the numerical solution of Kepler's equation than I would want to read. I have not examined any of it (almost) and strongly suggest that you follow my lead." In fact, Taff was commenting on quite modern numerical work on KE we'll be describing in Chapter 8.

Tables and Approximation Formulas

In 1855 W. Lehmann wrote a note to the editor of the influential journal, *Astronomische Nachrichten,* [Lehmann,1855], to outline his ideas of what were the most important problems for astronomy to solve. The main problem, in his view, was to achieve a "definitive solution" of the Kepler problem. This solution would be a table with entries E of substantial accuracy for a grid of pairs (e, M) fine enough that interpolation in e and M would yield accurate solutions for KE in all cases. The second problem he saw as the need for a complete (mathematical) study of interpolation which would produce a definitive routine to interpolate table readings to a standard all could accept.

From about 1820 up to the present *Astronomische Nachrichten* has been an important astronomical journal. In our Bibliography we have listed over 50 citations from it which bear directly on KE. Although it is not possible to claim that Lehmann's call for a solution of KE actually caused an outpouring of responses, we detail over 50 citations dated after 1855 devoted to carrying out in some sense the program he called for. Of these, 20 are concerned with plans for, or actual construction of, analogue devices for solving KE, and eighteen appeared in *Astronomische Nachrichten.* Lehmann himself devoted much of his work between 1855 and 1858 to an extended project of creating astronomical tables, the reporting of which involved 18 separate entries in Volumes 44–47 of this journal (most of which we do not include in our Bibliography).

One of the very prolific authors of the time who studied KE seriously was A. DeGasparis. In several papers of 1857 he described an analogue device like a slide rule to find solutions of KE, [DeGasparis,1857a,b,c,d,g]. The device worked this way:

If $h = 1°$ and

$$d_j = \log\left(\frac{h}{\sin jh}\right)$$

for $j = 1, 2, 3, \ldots$, prepare on two strips scales with equally spaced tics marked as in Figure 7.1.

1	d_1	d_2	d_3
	h	2h	3h

2			
	log0	log1	log2

Figure 7.1

Given M, choose integer p so that ph is closest to M, and place Rule 1 above Rule 2 so that position ph on Rule 1 lies over position $\log 0$ on Rule 2. In this configuration each d_j on Rule 1 lies above some $\log k$ on Rule 2, and one chooses k so that $d_j + \log k$ is closest to $\log e$. Then kh will approximate E in KE to the nearest degree.

To produce a solution of KE accurate to within one second of arc, De-Gasparis would first find integers p_1, p_2 so that $p_1 h \leq M \leq p_2 h$, and use his rules to obtain $k_1 h$ and $k_2 h$. By linear interpolation he would determine E_1 so that

$$\frac{E_1 - k_1 h}{(k_2 - k_1)h} = \frac{M - p_1 h}{(p_2 - p_1)h}$$

and then take

$$E = E_1 + \frac{M - E_1 + e \sin E_1}{1 - e \sin E_1}.$$

Of the 48 columns in the paper [DeGasparis,1857a], 47 are devoted to a table of solutions of KE obtained in this manner. DeGasparis noted that a similar device might be used as well to solve

$$M = E \pm e\phi(E)$$

for E with any tabulated function ϕ, with scale h being arbitrarily chosen.

In 1862, 1877, and even 1881, DeGasparis continued to publish refinements and variations on this approach to KE involving computation by logarithms and a single, Newton-like correction.

Between 1855 and 1890, when the extensive tables of J.J. Astrand were published, quite a variety of practical advice for solving KE was given, and various tables of solutions appeared.

At roughly the same time as DeGasparis, J.P. Wolfers proposed an iteration scheme, [Wolfers,1862], which also involved some table-constructing. Given M and e, Wolfers used E_1, given by

$$\tan E_1 = \frac{\sin M}{\cos M - e} \tag{7.1}$$

as a first approximation of E, with subsequent refinements defined by

$$\left\{ \begin{array}{l} \sigma_n = (E_n - M) - \sin(E_n - M) \\[2ex] k_n = \dfrac{-\sigma_n \cos E_n}{\cos M - e} \\[2ex] E_{n+1} = E_n + k_n. \end{array} \right\}$$

The initial estimate E_1 of Wolfers had been considered by W. Wallace, [Wallace,1835,1836], along with a rediscovery of Euler's method. Even earlier, [Trembley,1782], it had been used by Trembley, who recognized that it is a poor approximate solution if e is not small. Wolfers had used this method on Encke's Comet ($e = 0.8474362$), iterating 13 times before getting the accuracy equivalent to 4 iterations by Gauss' method. Part of Wolfers' effort was the preparation of an auxiliary table for

$$\sigma(x) = x - \sin x$$

for arguments in the range $0°$ to $18°$.

W. Doberck, [Doberck,1878], published tables to determine E from (e, M) and M from (e, E) in *Astronomische Nachrichten*, but he gave no clue as to how he obtained them. Later A. Seydler, [Seydler,1888], used Doberck's tables as first approximations of E and created three auxiliary tables to give corrected values. Which table would be used to correct depended in a complicated way on the size of e and the size of Doberck's table reading.

J.F. Encke, [Encke,1850], used a scheme to approximate E which is essentially two iterations of Newton's method followed by an interpolation. Given

e and M, he calculated

$$\left.\begin{cases} \tan E_1 = \dfrac{e\sin M}{1 - e\cos M} \\[2ex] M_1 = E_1 - e\sin E_1 \\[2ex] \tan E_2 = \dfrac{e\sin(M - M_1)}{1 - e\cos(M - M_1)} \\[2ex] M_2 = E_2 - e\sin E_2. \end{cases}\right\} \qquad (7.2)$$

Then with

$$a = \frac{M_2 - M_1}{M_1}$$

$$R = M_2 + \left(\frac{a}{1 - a}\right)(M_2 - M_1)$$

$$S = E_2 + \left(\frac{a}{1 - a}\right)(E_2 - E_1)$$

he took $E = M - R + S$, claiming accuracy to $1.5''$.

Later N. Herz, [Herz,1881], also used (7.2) for a starting approximation E_1, with

$$c = \frac{1}{\mathrm{arc}\,1''}$$

and $N = \sin(\arctan E_1)$, and set

$$E = M + cN - \frac{c\cot M}{6}N^4 + cN^5 + \frac{11}{12}c(\cot M)N^6.$$

This approximation of E appeared in Th.v. Oppolzer's influential *Lehrbuch zur Bestimmung der Kometen und Planeten*, [Oppolzer,1882].

Oppolzer himself used methods found in [Grunert,1856] and [Zenger,1882] to produce an extensive table of solutions of KE, [Oppolzer,1885]. In hopes of improving on Oppolzer, A. Marth created auxiliary tables in [Marth,1890b,c] to be used as corrections to Oppolzer's entries. One table calculated

$$L = L(e, E) = \frac{e\sin E}{\sin(e\sin E)}.$$

Taking Oppolzer's tabulated values for E_1, Marth would read $L_1 = L(e, E_1)$ and use

$$\tan E = \frac{L_1\sin M}{L_1\cos M - e}$$

to determine E.

The Astrand tables, [Astrand,1890], came as close as any to what Lehmann had called for. They were designed with e-increments of 0.01 and a grid of M-values fine enough that with errors in M no greater than $0.01''$, interpolation by the method of *regula falsi* would produce E-values accurate to $0.01°$. Since this interpolation required a division by $1 - e\cos E$, when this quantity was small an auxiliary table was to be invoked which implicitly called on a parabolic approximation (and Barker's equation.) Although Astrand's tables did not end efforts at table-building for KE, they were the last major effort to publish complete solutions. Between 1900 and 1936 a variety of "solution-of-KE-by-auxiliary-table" enterprises were proposed and tested by example, but none gained much widespread use.

B. Viaro, [Viaro,1920], conceived a program based on the interpolation formula of Bessel which was applied to auxiliary tables of (E, M)-pairs and first and second differences in M-values. Viaro's scheme was both odd and interesting. Because Viaro's program required separate auxiliary tables for each value of e, it is not clear that Viaro ever intended it to be implemented. Were such tables available, the accuracy of Viaro's results would make his scheme as successful and convenient as any which appeared in the period 1860–1890; had his work been done in 1880 instead of 1920, it might have earned considerable attention.

Also interesting was Viaro's use of Bessel's interpolation formula. Bessel interpolation had considerable play before the age of computers but it is not much favored or recognized by numerical analysts today. (In studying Viaro's paper, belatedly I came to realize that in a pre-1960 numerical analysis course I had actually studied and partially mastered Bessel interpolation but had never encountered it since. Mature readers might share my lament: if only I knew everything I once studied!)

W.S. Aldis, [Aldis,1902], presented another interesting auxiliary table for KE. If $e = \sin\phi$,

$$M = E - e\sin E = E - \sin\phi\sin E$$
$$= \frac{1}{2}(\phi + E + \cos(\phi + E)) - \frac{1}{2}(\phi - E + \cos(\phi - E)).$$

With

$$C(\theta) = \frac{1}{2}(\theta + \cos\theta)$$

we have

$$M = C(\phi + E) - C(\phi - E).$$

Aldis tabulated $C(\theta)$ in seconds for all θ, $-90° < \theta < 90°$, in increments of

$1'$. Given M and $\phi = \sin^{-1} e$, he let

$$d'' = \frac{C(\phi + 1') - C(\phi - 1')}{2}$$

and, expressing M in minutes, he took

$$E_1 = \left(\frac{M}{2d}\right)'.$$

Linear interpolation on $C(\phi + E_1)$, $C(\phi - E_1)$ will give an improved value for E. Among several clever uses of his table, Aldis described a way to estimate $x - \sin x$ when x is small.

The last occurrence of an auxiliary table for KE we have found was done by G. Draper, [Draper,1936], who calculated first, second, and third differences of the quantities

$$\frac{B_0}{e} = \sin E_0$$

in degrees when E_0 is an initial estimate of E in degrees, denoting them by

$$\frac{B_0'}{e}, \quad \frac{B_0''}{e}, \quad \frac{B_0'''}{e}.$$

The table readings were used with the following iteration scheme: with $M_0 = E_0 - e \sin E_0$ and

$$\Delta_0 = \frac{M - M_0}{1 - B_0'}$$

let

$$\left\{\begin{array}{l} \Delta_{n+1} = \dfrac{M - M_0}{1 - B_0' + \Delta_n B_0'' + \Delta_n^2 B_0'''} \\[2ex] E_{n+1} = E_0 + \Delta_n \\[2ex] n = 0, 1, 2, \dots \, . \end{array}\right\}$$

Draper claimed that for any $e < 1$, E_2 in degrees would be accurate to $(2 \times 10^{-7})°$.

To indicate more completely the extent of the work done on KE in the period 1900–1930, we present with few details four approximate solutions. While their derivations are not particularly interesting, they represent results which both recall the era of the 1860's when such approximations were sought and preview the surprising amount of attention given to robust and exceedingly accurate solutions of KE since 1968. We are not able to find that any of them were much used, although the last of them has received a recent brief mention.

In each case we presume e and M are given.

1. [Witt,1906]: Define E_0 by

$$\tan E_0 = \frac{e \sin M}{1 - e \cos M}$$

and take

$$E = M + \arctan \left\{ \frac{1 + \dfrac{2}{3} \left| \dfrac{\ln \sec E_0}{1 - e \cos M} \right|}{1 + \left| \dfrac{\ln \sec E_0}{1 - e \cos M} \right|} \right\}.$$

2. [Hartmann,1918]: Define

$$E_0 = \frac{\sin M}{\cos M - e}$$

and take

$$E = E_0, \quad \text{if} \quad e < 0.25$$

$$E = \arctan \left\{ \frac{\sin M}{\cos M - e + \dfrac{e}{2} \left(\dfrac{E_0 - M}{100} \right)^2} \right\}, \quad \text{if} \quad e > 0.25.$$

3. [Wilkens,1918]: For any initial guess E_0 take

$$E = M + \arctan \left\{ \frac{e \sin M}{\left[1 - \frac{1}{15}(E_0 - M)^2 \right]^{-5/2} - e \cos M} \right\}.$$

4. [Brown,1931]: Take

$$E = M + \arcsin \left\{ \frac{e \sin M}{\sqrt{1 + e^2 - 2e \cos M}} \right\}. \tag{7.3}$$

Analogue Devices

Quite a few attempts have been made, or at least conceived, to describe a solution of KE by analogue means—making a geometric measurement, reading from a scale, or estimating a point of intersection of curves. No classification we tried to help describe them seemed right; we chose to arrange them according to their mathematical basis rather than how the scheme is implemented. From this aspect the devices may be sorted into five categories.

1. **Cassini's approximation.** In Chapter 2 we described the 1856 device of Carlini based on the Cassini approximation. We find that Y. Ueta in 1927 gave a plan for a mechanical device, [Ueta,1927], to solve KE in precisely the same manner—rotating a circle until a straight line and a radius are perpendicular. Ueta's device was less elaborate than Carlini's.

2. **Curves of sines.** Also in Chapter 2 we described the simple "curve of sines" solution proposed by Adams in 1849. R.A. Proctor, whose role in changing 1870 attitudes about the origin of comets is described in [Bailey,1990, pp. 134–135], delivered a paper at a meeting of the Royal Astronomical Society in 1873, [Proctor,1873], which described an elaborate method to solve KE by drawing ellipses and a family of sine curves. Adams himself was presiding at the meeting, and he remarked after Proctor's talk that Proctor's method was actually equivalent to his own solution by a single "curve of sines." In 1877 C. Pritchard, [Pritchard,1877], presented two protractor-like devices to read off solutions of KE conveniently. Both were based on the curve of sines, and the second was a simplification of Proctor's scheme.

   In 1895, F.R. Moulton, [Moulton,1895], and T.J.J. See, [See,1895,1896], both described the same simple device to solve KE with the curve of sines: given e, prepare a right triangle with vertex angle ψ, where

   $$\tan \psi = \frac{1}{e}$$

   so as to locate points P and E as in Figure 2.11. If one cares to solve KE repeatedly for the same e and varying M, this would be a convenient way to approximate E. Both Moulton, who was concerned with comets, and See, whose interest was the study of binary stars, had considerable need to solve KE repeatedly.

   F. Dolainski, [Dolainski,1913], took the trouble to describe the curve of sines method once again, claiming a novelty for his presentation which is not apparent. Earlier we noted that Waterston in 1850 and Dubois in 1863 rediscovered Adams' method. Some thirty years after Dubois, K. Zelbr submitted a short paper, [Zelbr,1895], whose only purpose was to state that Dubois' work had been preceded by Waterston. It is difficult to see what motivated Zelbr to write such a note.

3. **Solution by cycloid.** Figure 2.10 and its accompanying description outline the basis for a feasible device to solve KE quickly: if position D can be easily adjusted so that $CD = e\,CO$ when the circle is rolled to the right until D is positioned directly over $H = M$, then C will be positioned directly over $F = E$.

Several were moved to describe such a device. O. Callendreau, in a footnote to the last sentence of the paper [Callendreau,1879], referred to one such as a "rapporteur." A. Rambaut actually described a device in [Rambaut,1890] but conceded that rolling without slippage is hard to do.

H. Plummer, [Plummer,1896], also described a cycloid device which was simple but not very easy to use. E. Wilczynski reinvented the method in [Wilczynski,1912] and constructed a device as a consequence of studying Newton, evidently with no awareness of the efforts of others.

In 1906 Rambaut used a section of the involute of a circle to fabricate a KE-solving device which also had some slippage problems, [Rambaut,1906]. In [Rambaut,1913] he improved the device by using two protractors. In this paper he included a proper and fairly complete record of the uses of the cycloid to solve KE, with references from Newton to Wilczynski.

The latest mechanical KE-solver we find described is that of A. Dazew, [Dazew,1934]. Based on the method of cycloids, it featured a working model based on a circle of 21 cm. and a thumbscrew mechanism with elaborate scales. Dazew evidently built his device with wood, claiming accuracy of $5'$, and later remade it of metal to improve accuracy to $1'$.

4. **Use of a slide rule.** In the papers [Mache,1910] and [Levin,1922] description was made to use a slide rule to approximate E to an accuracy of two or three minutes. Both attempted to improve their slide rule readings by linear interpolation.

In the earlier effort, H. Mache reports that he was communicating an approximation he found "in the writings of my recently-deceased father." The description and examples are hard to follow, so it isn't clear how well Mache understood what he was writing.

Actually for both Mache and Levin the role of the slide rule was almost incidental, it being used only to obtain a first approximation E_1 given M and e. Both Mache and Levin would then calculate $M_1 = E_1 - e \sin E_1$ and use

$$E_2 = E_1 + \frac{M - M_1}{1 - e \cos E_1}. \tag{7.4}$$

With $f(E) = E - e \sin E$, the linear approximation of f at E_1 is

$$f(E) \approx f(E_1) + f'(E_1)(E - E_1)$$

or

$$M \approx M_1 + (1 - e \cos E_1)(E - E_1)$$

and (7.4) is the resulting approximation of E.

Were one simply to take $E_1 = M$, and ignore the slide rule, it would be the case that $M_1 = M - e\sin M$ and (7.4) would read

$$E_2 = M + \frac{e\sin M}{1 - e\cos M}. \tag{7.5}$$

The approximation (7.5) is equivalent to using one iteration of Newton's method in solving

$$g(E) = E - e\sin E - M = 0$$

for E with initial guess $E_1 = M$. When e is quite small, say $0 < e < 0.1$, E_2 in (7.5) is quite a good approximation of E. (In fact, $|E - E_2| < 5.6 \times 10^{-4}$.) We discussed in Chapter 4 how sensitive Newton's method can be to the gap between the true solution and the initial guess. When $e > 0.1$, the initial guess $E_1 = M$ may require more iterations than one cares to perform by hand or by calculator.

Given M and e, quantity (7.5) supplies an initial value for E from which Newton's method converges quickly for a broad range of (e, M)-pairs. A modern appearance of this observation is found in [Duffet-Smith,1981, Figure 28] where are drawn graphs of

$$\Delta_e(M) = \frac{e\sin M}{1 - e\cos M}$$

for several values of e, which are called "Kepler's graphs." On the curve corresponding to a value of e, the ordinate $\Delta_e(M)$ corresponding to abscissa M is the correction of M yielding E_2 in (7.5), to be used either as an approximation of E or as the initial guess for Newton's method. (Chapter 8 chronicles modern computing games and the search for an "optimal" initial guess in Newton's method applied to KE.)

5. **Nomographs and other graphical solutions.** A nomograph is defined in *The American Heritage Dictionary of the English Language* to be:

> A graph consisting of three coplanar curves, usually parallel straight lines, each graduated for a different variable, so that a straight line cutting all three curves intersects the related values of each variable. [Morris,1981, p. 892].

A nomograph can be a convenient way to represent in two dimensions the coordinates (x, y, z) of all solutions of an equation $f(x, y, z) = 0$

when only solutions of low accuracy are required. In an age when we can have electronically highly accurate numerical data or its graphical display almost instantly from a personal computer, it is hard to assign much importance to nomographs now. Nomography did have its proponents as a serious approach to dealing with equations some years ago. A rational treatment of the subject was given in [Brodetsky,1938], and a nice treatment of nomographs for astronomical problems appeared in [Wood,1946].

A nomograph for KE would display three curves scaled for the variables e, M, and E so that any straight line traversing the curves relates values of e, M, E which satisfy $E - e \sin E - M = 0$. The nomograph only presents the solutions, the actual solving having been addressed in marking the scales on the curves, but it's convenient to commit the small abuse of language in the phrase "nomographic solution of KE." We can describe two different graphical displays of solutions of KE which incorporate some of the features of a nomograph.

R. Radau, [Radau,1884], used a graphical device for KE which has some features of a nomograph. On a rectangle of convenient size with vertices as marked in Figure 7.2, mark the vertical scale in degrees and the horizontal scale in increments of 0.01. For each value of E in degrees draw a line from $(0, E^\circ)$ to $(1, (E - \sin E)^\circ)$. Each line so drawn is an "E-level curve" in the rectangle, and for each x, $0 < x < 1$, the point in the rectangle with abscissa x has coordinates $(x, (E - x \sin E)^\circ)$.

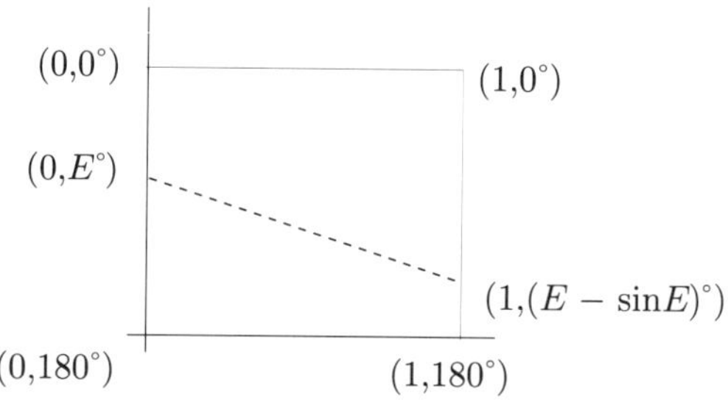

Figure 7.2

Given e and M, one locates the point in the rectangle with coordinates (e, M°), notes visually the "E-level curve" which passes nearest to (e, M°), and follows this level to the left edge of the rectangle to read E to the nearest degree. It was Radau's intent to use this value as an initial value to be improved by Newton's method. With an initial guess within $1^\circ = 0.018$ radians, this would give reasonably effective results. Radau's scheme was described also by E. Le Grand Roy, [Le Grand Roy,1903].

The nomograph we describe next appeared several times in the literature. The first appearance we find is by M. D'Ocagne, [D'Ocagne,1894], and the slight variation of it we treat appeared in [Fry,1916], [Rand,1942], [Jensen,1962], and [Duffet-Smith,1981].

In Figure 7.3 the left vertical line corresponds to e with a scale from 0 to 1 marked with increments of 0.01. The right vertical line measures M from 0 to π in increments of 0.01π. The curve attached to the M-scale and lying to its left is the graph of

$$h(M) = \frac{\sin M}{1 - \left(\frac{M}{\pi}\right)\cos M}$$

where M is marked vertically and $h(M)$ is plotted to the left.

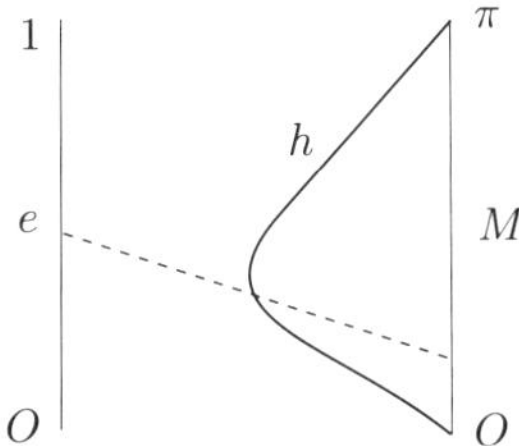

Figure 7.3

On the h-curve, tics would mark $h(M)$-values corresponding to increments of 0.01π in M.

Given e and M, a line from the e-scale to the M-scale intersects the h-curve at a point with some distance h^* to the left of the M-scale, and one would take $E = M + h^*$. In fact, this value of E represents the approximation (7.4).

The same graph will treat values of M, $\pi \leq M \leq 2\pi$, by connecting e on the left with $\pi - M$ on the right and, reading h^* from the h-curve, taking $E = M - h^*$. D'Ocagne's 1894 version incorporated all values of M from 0 to 2π. Fry's 1916 version added to Figure 7.3 extra scales so that one might read all the quantities E, r/a, and v.

We have looked at one other nomograph for KE devised by V. Laska, [Laska,1914], for cases when $e < 0.25$. Since each attempt to determine E from e and M with it requires three nontrivial auxiliary calculations and the drawing of two lines, it is quite a complicated scheme and we won't try to describe it.

Chapter 8

Modern Treatments of KE

Buried as we are in the Kepler problem, an old narrow problem well worked-over for many decades, our treatment of the period 1930–1990 can be relatively insulated from many of the immense and rapid changes which are evident in the study of almost any area of science for this time. Between 1930 and 1950 we find relatively little work on KE, most of it derivative to earlier efforts but beginning to take small advantage of calculating machinery.

Both sciences and scientists underwent great changes before, during, and after the years of World War II. With regard to our narrow direction only, developments leading to the realization of orbital satellites recharged celestial mechanics and problems of orbit determination, and the rise and growing momentum of computers began to transform scientific computation.

With new instruments, both optical and radio, observational astronomy was entering a time of great ferment, and Oort's fresh thinking about the origin of comets in 1950 was attracting new ideas and workers. By the early 1960's enough had been accomplished in all these areas to foster a rush of attention to KE which continues to this moment.

Our treatment is in four parts. The first covers 1930–1950, the second describes some aspects of universal formulations of KE, while the third describes extensive work on KE in the spirit of numerical experiments. In the fourth part we gather a few techniques for solving transcendental equations for which KE was used as an illustrious test case.

The Period 1930–1950

In previous chapters we have described parts of the work on KE published in the 1930's which had the flavor of much older efforts and which seemed to be of small value and little contemporary interest.

We can summarize briefly a few computational efforts of the time. In the same issue of the *Revista Astronomica of Buenos Aires* which contained [Moulton,1930], a translation of Moulton's 1895 paper rehashing the method of curve of sines, J. Bobone, [Bobone,1930], described a method to solve KE consisting of the following two steps:

(a) Let $E_0 = M + x$ so that

$$x = e \sin M \cos x + e \cos M \sin x.$$

Using only terms through x^2 in the power series for $\cos x$, $\sin x$, take the positive solution x of

$$\frac{1}{2}x^2 + \left(\frac{1 - e \cos M}{e \cos M} \right) x - 1 = 0.$$

(b) Now let $E = E_0 + y$ and similarly produce a quadratic equation in y. Bobone's choice of a solution y was predicated on the sign of the quantity

$$\frac{2(M - E_0 + e \sin E_0)}{e \sin E_0}$$

and E was thus obtained by two quadratic approximations.

The papers [Innes,1932,1933] and [Moller,1933,1940] evidence some interest in KE for large values of e, up to and including 1. While the Innes papers showed nothing new, the papers of Moller were intended to have the help of machine calculations. Moller started with changes of variables introduced in [Andoyer,1918] to change KE into a form more tractable when e is near 1. If $S(E) = E - e \sin E$, KE may be written

$$M - (1 - e) \sin E = S(E). \tag{8.1}$$

Moller prepared a table of $\sin E$ and $S(E)$ values. Given M and e, he proposed calculating $M - (1 - e) \sin E$ repeatedly for different values of E until it matched $S(E)$ closely. In this way he claimed to achieve very accurate values of E for e-values in the range $0.7 < e < 0.95$. Later S. Herrick, [Herrick,1945], made a slight modification to [Moller,1933] for the cases e near 1.

An interesting and novel approach to solving KE numerically was described by R. Langebartel, [Langebartel,1948]. In 1942 there had appeared a paper, [Agnew,1942], which extended a 1917 result by E. Whittaker expressing the minimum modulus root of a polynomial as a limit of ratios computed by determinants. The theorem Langebartel adapted to KE proved that if $z = r$ is the simple zero of minimum modulus of a function

$$f(z) = -1 + \sum_{j=1}^{\infty} a_j z^j$$

analytic in some disk $|z| < R$ with $R > r$, then

$$\lim_{n \to \infty} \frac{A_{n-1}}{A_n} = r$$

where

$$A_0 = 1, A_n = \sum_{j=1}^{n} a_j A_{n-j} \tag{8.2}$$

$$n = 1, 2, 3, \ldots$$

With

$$f(E) = \frac{E}{M} - \frac{e \sin E}{M} - 1$$

and

$$-1 + \left(\frac{1-e}{M}\right) + \frac{e}{3!M} E^3 - \frac{e}{5!M} E^5 + \cdots = 0 \tag{8.3}$$

the minimum modulus zero of f is the unique solution of KE. In this case

$$a_1 = \frac{1-e}{M}$$

$$a_{2k} = 0$$

$$a_{2k+1} = \frac{(-1)^{k+1} e}{(2k+1)!M}$$

$$k = 1, 2, 3, \ldots$$

The calculation of the quantities A_1, A_2, A_3, $\ldots$ in (8.2) is recursive and increasingly costly, but Langebartel found that, with the aid of some simple changes of variables based on the size of M, the ratio A_5/A_6 would suffice to obtain E accurate to 5 places. Langebartel demonstrated that the method is feasible for machine calculations, with one of the advantages being that no evaluations of trigonometric functions are required. An exposition of Langebartel's method for KE with computations comparing its accuracy to Newton's method was given recently in [Dass,1990].

In addition to an extensive bibliography of papers on KE, H. Wood in [Wood,1950a,b] gave some treatment of numerical calculation by machines of the time.

"Universal" Forms for KE

By 1960 problems of celestial mechanics and orbit determination were being reformulated, and indeed new sciences like "astronautics" and "astro-dynamics" were building to study them. As E. Pitkin wrote, [Pitkin,1965a],

The recent introduction of rocket propulsors for space flight has presented possibilities of encountering orbits whose character changes significantly during the time of interest. For instance, a low-acceleration escape trajectory might begin in a nearly circular orbit and terminate on a hyperbolic path. This aspect, as well as the nearly universal use today of electronic computers for orbit computation and the attendant desire to simplify programming, motivate one to employ a formulation free from possibilities of indeterminateness, no matter what form the orbit takes, elliptic, parabolic, hyperbolic, or possibly a circular or linear limit.

At this time there were efforts in several directions to reformulate laws of orbital motion, and in particular KE, in "universal" variables which incorporate all eccentricities and allow rapid, accurate computation of positions and velocities. Several approaches, with different choices of universal variables, are to be found in [Battin,1964], [Herrick,1971], [Stumpff,1968a], [Gooding,1987b], and most recently [Sarnecki,1988].

Since it has been our plan to treat KE only for elliptic orbits, and our discussion until now has not required any mention of a form of KE for hyperbolic orbits, we have made no prior preparation for a context in which to describe a universal form of KE. We'll describe such a form, as given in [Battin,1964], very briefly here. More of the motivations and details are gathered in Appendix I.

Suppose, for any eccentricity $e > 0$, an object following a path of eccentricity e is at time $t = 0$ in position R_0 with velocity V_0, relative to a focus, with eccentric anomaly E_0. A universal Kepler's equation (UKE) is given by

$$\sqrt{\mu}\,t = \frac{R_0 \bullet V_0}{\sqrt{\mu}} x^2 C(\alpha x^2) + (1 - |R_0|\alpha)x^3 S(\alpha x^2) + |R_0|x \qquad \text{(UKE)}$$

where μ = gravitational constant and $\alpha = 1/a$ and the functions C, S are defined to be

$$C(x) = \sum_{n=1}^{\infty} \frac{(-1)^{n-1} x^{n-1}}{(2n)!}$$

$$S(x) = \sum_{n=1}^{\infty} \frac{(-1)^{n-1} x^{n-1}}{(2n+1)!}.$$

In terms of elementary functions

$$
C_{(}x) = \begin{cases} \dfrac{1 - \cos \sqrt{x}}{x}, & x > 0 \\[2em] \dfrac{\cosh \sqrt{-x} - 1}{-x}, & x < 0 \end{cases}
$$

and

$$
S(x) = \begin{cases} \dfrac{\sqrt{x} - \sin \sqrt{x}}{(\sqrt{x})^3}, & x > 0 \\[2em] \dfrac{\sinh \sqrt{-x} - \sqrt{-x}}{(\sqrt{-x})^3}, & x < 0. \end{cases}
$$

When (UKE) is solved for x, the distance of the object from the focus is given by

$$
r(x) = |R_0| + (1 - \alpha|R_0|)x^2 C(\alpha x^2)
$$
$$
+ \frac{R_0 \bullet V_0}{\sqrt{\mu}} x\big(1 - \alpha x^2 S(\alpha x^2)\big).
$$

In the case where $0 < e < 1$, the elliptic case, variable x is given by

$$
x = \frac{E - E_0}{\alpha^{1/2}} = \sqrt{a}(E - E_0)
$$

while

$$
e \cos E_0 = 1 - \frac{|R_0|}{a}
$$
$$
e \sin E_0 = \frac{R_0 \bullet V_0}{\sqrt{\mu a}} \, .
$$

Substitution of these quantities in (UKE) will produce the more familiar form of KE with initial data $t = 0$, R_0, V_0, E_0,

$$
\sqrt{\frac{\mu}{a^3}}\, t = (E - E_0) - e(\sin E - \sin E_0) \tag{KE}
$$

in the following steps:

$$
\sqrt{\mu}t = a^{3/2} e \sin E_0 (E - E_0)^2 C\big((E - E_0)^2\big)
$$
$$
+ \left(1 - \frac{|R_0|}{a}\right) a^{3/2}(E - E_0)^3 S\big((E - E_0)^2\big)
$$

$$+a^{3/2}(1 - e\cos E_0)(E - E_0)$$

$$\sqrt{\frac{\mu}{a^3}}\,t = e\sin E_0 \sum_{n=1}^{\infty} \frac{(-1)^{n-1}(E - E_0)^{2n}}{(2n)!}$$

$$+e\cos E_0 \sum_{n=1}^{\infty} \frac{(-1)^{n-1}(E - E_0)^{2n+1}}{(2n+1)!}$$

$$+(1 - e\cos E_0)(E - E_0)$$

$$\sqrt{\frac{\mu}{a^3}}\,t = e\sin E_0\big(1 - \cos(E - E_0)\big)$$

$$+e\cos E_0\big((E - E_0) - \sin(E - E_0)\big)$$
$$+(1 - e\cos E_0)(E - E_0)$$
$$= e\sin E_0 - e\sin E_0 \cos(E - E_0)$$
$$-e\cos E_0 \sin(E - E_0) + (E - E_0)$$
$$= (E - E_0) - e(\sin E - \sin E_0).$$

In Appendix I the relation between (UKE) and KE in the parabolic and hyperbolic forms is detailed.

This abbreviated look at one form of a universal Kepler's equation conveys adequately the sophisticated nature of such treatments and will permit us to describe the variety of numerical experiments made with KE in recent years.

Numerical Experiments with KE

In solving KE by Newton's method for a given e and M, one uses a starting value E_0 dependent on e and M and iterates to obtain an approximate solution accurate to some prescribed tolerance. The number of iterations required to achieve this accuracy is a function of E_0, e, and M. As we mentioned in Chapter 4, E. Schubert in 1854 asked if one might find a choice for E_0 so that one iteration of Newton's method would achieve the desired accuracy for all e and M.

With the availability of wonderful computing machinery in recent years, questions like Schubert's have occurred to others and have moved them to subject KE and Newton's method to a variety of "numerical experiments."

Suppose f is a nice function for which $f(E) = 0$, but $f'(E) \neq 0$, and suppose with starting guess x_o we use Newton's method to generate a sequence

$$\{x_n\}_{n=1}^{\infty}$$

of iterations to approximate E. If for $n = 0, 1, 2, \ldots$, we call $\delta_n = x_n - E$, a simple argument based on Taylor's theorem will show that $x_n \to E$ as $n \to \infty$ and, for some constant K

$$\delta_{n+1} \leq \frac{K}{2} \left[\frac{f''(E)}{f'(E)} \right] \delta_n{}^2 \tag{8.4}$$

and as well

$$\delta_{n+1} \leq \frac{K \left[\frac{1}{2} \delta_0 \left\{ \frac{f''(E)}{f'(E)} \right\} \right]^{2^n}}{\left[\frac{f''(E)}{f'(E)} \right]}. \tag{8.5}$$

In KE with $0 < e < 1$, $f(x) = x - e \sin x - M$ so

$$\left| \frac{f''(E)}{f'(E)} \right| = \left| \frac{e \sin E}{1 - e \cos E} \right| \leq \frac{e}{1 - e}. \tag{8.6}$$

O.K. Smith, [Smith,1968], used (8.4) and (8.6) with KE to show that for all $e < 0.99$

$$|\delta_{n+1}| < 5 \times 10^{-9} \quad \text{when} \quad |\delta_n| < 10^{-5} \tag{8.7}$$

and for all $e < 1$

$$|\delta_{n+1}| < 10^{-8} \quad \text{when} \quad |\delta_n| < 10^{-4} \sqrt{\frac{1 - e}{e}}. \tag{8.8}$$

G. Smith, [Smith,1979], computed for many (e, M)-pairs the mean and standard deviation of the number of iterations required to solve KE to an accuracy of 5×10^{-8} for each of six starting values E_0. It was his finding that the starting value

$$E_0 = M + \frac{e \sin M}{1 - \sin(M + e) + \sin M}$$

produced the lowest mean and standard deviation for the iteration number over the (e, M)-pairs he used. With special attention to the region $\{(e, M) : e \geq 0.95, M \leq 0.4\}$, which is more difficult for Newton's method, the simpler starting value $E_0 = M + e$ was best of those he used. Recently E. Taylor, [Taylor,1989], plotted the iteration numbers $N = N(e, M)$ as surfaces for each of Smith's starting values.

Almost simultaneously with G. Smith, E. Ng in [Ng,1979] made a similar experiment with the iteration function

$$x - \frac{f(x)}{f'(x)} - \frac{1}{2} \frac{[f(x)]^2 f''(x)}{[f'(x)]^3} \tag{8.9}$$

which is known as Schröder's method or Tchebycheff's formula. Ng used for E_0 three polynomial functions of e and M according as (e, M) lay in one of the regions

$$M_0 = \tfrac{\pi}{2} - e$$

$$0.7 < e < 1, \quad 0 \le M \le M_0:$$

$$E_0 = 1 - (1 - \tfrac{M}{M_0})^4$$

$$0 \le e < 0.7, \quad 0 \le M \le M_0:$$

$$E_0 = 1 - \left(1 - \frac{M}{M_0}\right)^2 \left(1 + \frac{2M}{M_0} - \frac{M}{1-e}\right)$$

$$M_0 \le M \le \pi:$$

$$E_0 = 1 - \left(1 - \frac{\pi - M}{\pi - M_0}\right)^2 \left[1 + \frac{2(\pi - M)}{\pi - M_0} - \frac{\pi - M}{1+e}\right].$$

With this scheme, Ng found that accuracy of 10^{-8} would be achieved in one iteration over 42% of the total (e, M)-region, and in two iterations over 98.8% of the region.

With an eye to improving computational efficiency, J. Prussing, [Prussing,1977,1979], gave some attention to bounds for the solution of KE in both the usual and universal forms described earlier. R. Broucke, [Broucke,1980], sharpened these bounds and formulated from them a cubic polynomial

$$E_0 = \left(\frac{2e^2}{1 - e^2}\right) \frac{M^3}{\pi^2} - \frac{e(1 + 3e)}{1 - e^2} \frac{M^2}{\pi} + \frac{M}{1 - e} \tag{8.10}$$

to use as an initial value. Broucke used (8.9) and computed to an accuracy of 10^{-12} with (8.10) and various other starting values, including those of Ng and G. Smith. He also calculated mean and standard deviation of the number of iterations and found those of Ng and G. Smith to be the smallest.

Even more than Newton's method, generalizations of it like (8.9) are known to be extremely sensitive to initial values. An interesting feature of Broucke's paper was a short list of specific (e, M, E_0)-values for which (8.9) produced very slow convergence, slow convergence to stable fixed points of various orders, and convergence to so-called "strange attractors." For example, the points (0.6822,1.208,0.45) and (0.8802,0.242,0.23) with (8.9) produce iterates which may vary erratically between narrow bounds with no discernible regularity or convergence. (For those who care to reproduce Broucke's results, [Broucke,1980, p. 259, Equation (10)] should be corrected to (8.9).) This behavior is known now to appear quite commonly in many iteration schemes, and the area of choatic dynamics has experienced enormous growth in exhibiting and explaining behavior in deterministic systems. A look at the hazards of using Newton's method without choosing initial values carefully appears in M. Barnsley, *Fractals Everywhere,* Academic, New York, 1988, pp. 80–90.

M. Bergam and J. Prussing, [Bergam,1982], made extensive numerical tests of both Newton's method and (8.9) in the context of Battin's (UKE), comparing the results of Ng, G. Smith and Broucke with three new starting values of their own.

The first, x_m, was the average of refinements made by Bergam, [Bergam,1980], to the upper and lower bounds for E obtained in [Prussing,1979]. The second, x_p, was a parabolic approximation obtained by solving exactly for the eccentric anomaly along a parabolic path with the same perihelion and the same R_0 and V_0 at time 0 as the actual orbit. The third and most effective starting value, x_c, was the solution of the following cubic equation obtained from (UKE) by using x_m in the arguments of C and S:

$$\sqrt{\mu}t = \frac{R_0 \bullet V_0}{\sqrt{\mu}} x^2 C(\alpha x_m{}^2)$$

$$+(1 - |R_0|\alpha)x^3 S(\alpha x_m{}^2)$$

$$+|R_0|x_m.$$

While not the best for every choice of e or choice of iteration scheme, x_c seemed the most robust over all values of c from 0 to 4. Bergam and Prussing alluded to some discrete instances of chaotic behavior like those Broucke saw but gave no specific examples.

Robust as x_c may be, a certain overhead cost is attached to obtaining it as a starting value. A different and totally empirical scheme to determine a starting value was undertaken by B. Sheela, [Sheela,1982].

With a one-time computational cost, Sheela determined a 3×3 matrix of constants, A, so that

$$E_0(e, M) = (e^2, e, 1)A(M^2, M, 1)^T$$

when used as a starting value in Newton's method achieves accuracy of 10^{-13} for all $e < 1$ in a number of iterations whose mean and standard deviation are less than those reported by Broucke. In the eight special cases where Broucke reported unusual behavior, Sheela's starting value led to solutions of this accuracy in three iterations.

Here is how Sheela obtained matrix A.

(a) For each of 620 (e, M)-pairs, $\{(e_k, M_k) : k = 1, 2, \ldots, 620\}$ he used Broucke's starting value (8.10) to determine E_k accurate to 10^{-12}.

(b) With $S_k = (e_k^2, e_k, 1) A (M_k^2, M_k, 1)^T$ he let

$$F(A) = \sum_{k=1}^{620} (S_k - E_k)^2$$

and used a simplex method to determine entries for A which minimize $F(A)$.

Neither the choice of order 3 for A nor the number 620 were explained.

The challenge of achieving accurate solutions of KE by refinement of a starting value and a small number of iterations of Newton's method and some of its extensions as we have described it so far is summarized in several survey articles which appeared between 1983 and 1986. J. Danby and T. Burkhardt covered the elliptical case in [Danby,1983] and the hyperbolic and elliptical cases in [Burkhardt,1983]. R. Gooding and A. Odell made similar surveys in [Gooding,1985] and [Odell,1986]. To indicate the extent to which Schubert's original question had been answered by 1985, the computer program EKEPL2 of [Gooding,1985] will determine E for any e, $0 < e \leq 1$, and M in no more than two iterations so that both the error in E and in $|E - e \sin E - M|$ are less than 10^{-14}. Close as this is to Schubert's goal, there is a considerable amount of work we have not yet described.

R. Peters, [Peters,1984], wrote E in a series of powers of e whose coefficients are functions of a normalized variable f defined to be

$$f = \left\{ \begin{array}{ll} \dfrac{M}{\frac{\pi}{2} - e}, & 0 \leq M \leq \dfrac{\pi}{2} - e \\[3ex] \dfrac{M - \left(\frac{\pi}{2} - e\right)}{\frac{\pi}{2} + e}, & \dfrac{\pi}{2} - e \leq M \leq \pi. \end{array} \right\}$$

If

$$E = \sum_{n=1}^{\infty} k_n(f) e^n$$

and

$$S = \sin\left(\frac{f\pi}{2}\right), \quad C = \cos\left(\frac{f\pi}{2}\right), \quad G = S - f,$$

then

$$k_1(f) = S$$

$$k_2(f) = CG$$

$$k_3(f) = Ck_2(f) - \frac{1}{2}G^2 k_1(f)$$

$$k_4(f) = Ck_3(f) - Gk_1(f)k_2(f)$$
$$- \frac{1}{6}G^2 k_2(f)$$

$$k_5(f) = Ck_4(f) - Gk_1(f)k_3(f)$$
$$- \frac{1}{2}k_1(f)[k_2(f)]^2 - \frac{1}{2}G[k_2(f)]^2 + \frac{1}{24}G^4.$$

Peters used as his starting value

$$E_0 = k_1 e + k_2 e^2 + k_3 e^3 + k_4^* e^4 + k_5^* e^5$$

where k_4^*, k_5^* are only the first two terms of k_4, k_5, respectively, with iteration function (8.9). His findings were that his E_0 outperformed the starting values of G. Smith, Broucke, and Sheela in all these aspects:

(a) computing time;

(b) average and standard deviation of iteration numbers over 2000 (e, M)-pairs;

(c) maximum number of iteration numbers required in a single case up to $e = 0.88$.

Only for an e-value exceeding 0.888 did his procedure require as many as three iterations.

In a pair of papers, [Boltz,1986,1987], F. Boltz set out to improve on the starting value (8.10) of Broucke, which is a cubic polynomial in M, by creating a polynomial cubic in both e and M.

For each fixed e, Broucke's E_0 and the exact E solving KE are functions of M which at $M = 0$ and $M = \pi$ have the same value and the same first derivative value. Boltz intended to find as functions of e B_1, B_2, B_3, C_2, C_3 so that the equation

$$B_1 E + B_2 E^2 + B_3 E^3 = M + C_2 M^2 + C_3 M^3 \tag{8.11}$$

has a solution E_B as a function of M whose values and derivative values match those of E at $M = 0$, $\pi/2 - e$, and π for each e.

The actual B_1, B_2, B_3, C_2, C_3 Boltz determined are displayed in Appendix J. Boltz then solved (8.11) for E_B for many (e, M)-pairs and used E_B as the starting value with both the iteration function (8.9) and the iteration function

$$x - \frac{\dfrac{f(x)}{f'(x)}}{\left\{1 - \dfrac{f(x)f''(x)}{2[f'(x)]^2}\right\}} \tag{8.12}$$

to which (8.9) is an approximation. While not comparing his results directly to Broucke's and Sheela's by counting iterations, Boltz showed that in almost all elliptic cases his E_B gave a starting value closer to the solution E than Broucke's or Sheela's. In [Boltz,1987], Boltz looked at the hyperbolic form of KE in a similar manner.

In the methods of Ng, Peters, and Boltz, the special value of the mean anomaly

$$M = \frac{\pi}{2} - e$$

plays a key role in the design of the methods, and its explanation is interesting.

For any e, KE has the solution $E = \pi/2$ when $M = (\pi/2) - e$. In Kepler's method

$$\left\{ \begin{array}{c} E_0 = \frac{\pi}{2} - e \\ E_n = M + e\sin E_{n-1} \\ n = 1, 2, 3, \ldots \end{array} \right\}$$

one finds

$$\frac{\pi}{2} - E_1 < \frac{e^3}{2} \quad \text{and} \quad \frac{\pi}{2} - E_2 < \frac{e^7}{8}.$$

In short, Kepler's method converges to the solution most rapidly when E_0 is close to $(\pi/2) - e$. This outcome is shown most convincingly in [Meeus,1991, p.185, Figure 3].

B. Conway, [Conway,1986], had the idea to use a root-finding method of Laguerre on KE and subject it to extensive testing. Conway's results were attractive enough that J. Danby, [Danby,1987], experimented with them further and R. Gooding, [Gooding,1987a,b], also.

If $f(x)$ is a polynomial of degree N with only real roots, Laguerre's method to find roots of f uses the iteration function

$$x - \frac{Nf(x)}{\max|f'(x) \pm K_N|} \tag{8.13}$$

where

$$K_N = \sqrt{|(N-1)^2[f'(x)]^2 - N(N-1)f(x)f''(x)|}.$$

It is known that Laguerre's method converges for any starting value, so applying it to polynomials with only real roots lets one avoid some of the pitfalls of Newton's method and its extensions. In applying the Laguerre method for polynomials to the transcendental function $f(E) = E - e\sin E - M$, Conway was taking a gamble, of course, but the results were good. With $N = 5$ and $E_0 = M$, and iterating until

$$|f(E_n)| = |E_n - e \sin E_n - M| < 10^{-12}$$

he found solutions in 3 or 4 iterations for $0 \le e < 1$, $0 \le M < \pi$, except in the range $0.99 \le e$, $0 \le M \le 0.1\pi$, where 5 iterations sufficed.

The choice $N = 5$ was arbitrary, but the experience of an immense number of trials was that neither larger nor smaller values of N had much effect on the results in more than isolated cases. Conway got similar results in using this scheme on (UKE); for $e < 1$ he used a starting value which depends on e and the orbital period, while for $e > 1$ he used the value x_p of Bergam and Prussing as a starting value.

Danby, in experimenting further with Conway's program, found that simply replacing $E_0 = M$ by $E_0 = M + 0.85e$ reduced iterations by one, generally, and by two, in the more troublesome "large-e, small-M" region. This change, together with replacing $N = 5$ by

$$N = \frac{6 - 4X}{3 - 4X}, \quad X = \frac{f' f'''}{(f'')^2}$$

at each step, allowed some further improvement.

In comparison with the EKEPL2 program of Gooding and Odell, [Gooding,1985], Danby found that the following scheme would suffice to produce accurate solutions in two iterations for all elliptic cases:

Use the starting value

$$E_0 = \begin{cases} M + \left[(6M)^{1/3} - M\right] e^2, & M < 0.1 \\[2ex] M + 0.85e, & M \ge 0.1 \end{cases}$$

with the iteration function

$$x - \frac{f(x)}{\left[f'(x) - \dfrac{1}{2}\dfrac{f(x)f''(x)}{f'(x)} - \dfrac{1}{4}\dfrac{[f(x)]^2[f''(x)]^2}{[f'(x)]^3} - \dfrac{1}{6}f'(x)f'''(x) \right]}.$$

In [Gooding,1987a,b], Gooding improved on the program EKEPL2, writing programs: EKEPL to approximate E in robust fashion while also controlling the errors relative to E in the approximation; and SHKEPL to handle similarly the hyperbolic case.

Gooding also amplified Danby's finding that Conway's scheme is not optimal. In the region $0.999 \le e < 1$, Gooding found the choice $N = 3$ superior to Conway's $N = 5$. Moreover, Gooding found that it is generally preferable to use the limiting case, $N \to \infty$, in (8.13):

$$x - \frac{f(x)}{\sqrt{|[f'(x)]^2 - f(x)f''(x)|}}. \tag{8.14}$$

In the "difficult" region, $0.999 \le e < 1$, with $E_0 = M$, one iteration with (8.14) followed by iterations with (8.13) and $N = 3$ seemed to give best results.

As Gooding has noticed, from $E_0 = M$ one iteration by (8.14) produces an approximation similar to (7.3) of [Brown,1931], but Brown's approximation has no connection with Laguerre's method.

S. Mikkola, [Mikkola,1987], devised a relatively simple and robust empirical procedure to find E with relative error no greater than 10^{-15}. Since in our view it seems closest to the ideal solution of Schubert's 1854 question of all the results we've seen, we'll take some time to describe it.

With the auxiliary variable $s = \sin(E/3)$, KE becomes

$$3 \arcsin s - e(3s - 4s^3) = M$$

and the approximation

$$\arcsin s \approx s + \frac{1}{6}s^3$$

leads to an approximate KE

$$3(1 - e)s + \left(4e + \frac{1}{2}\right) s^3 = M. \tag{8.15}$$

Solving (8.15) for s and correcting the solution by the quantity

$$\frac{-0.078s^5}{1 + e},$$

empirically arrived at to improve the worst error associated with the case $M = \pi$, Mikkola set

$$w = s - \frac{0.078s^5}{1 + e}$$

and used the starting value

$$E_0 = M + e(3w - 4w^3)$$

which already enjoys maximum absolute relative error no greater than 0.002.

To improve on E_0, Mikkola took $f(x) = x - e\sin x - M$, evaluated f, f', f'', f''', and $f^{(4)}$ at E_0, and then calculated

$$\left\{\begin{aligned} u_1 &= -\frac{f}{f'} \\[2ex] u_2 &= -\frac{f}{f' + \frac{1}{2}f''u_1} \\[2ex] u_3 &= \frac{-f}{f' + \frac{1}{2}f''u_2 + \frac{1}{6}f'''u_2^2} \\[2ex] u_4 &= \frac{-f}{f' + \frac{1}{2}f''u_3 + \frac{1}{6}f'''u_3^2 + \frac{1}{24}f^{(4)}u_3^3} \\[2ex] E &= E_0 + u_4. \end{aligned}\right. \qquad (8.16)$$

The method requires solution of a cubic polynomial with no trigonometric functions and one-time evaluation of $\sin w$ and $\cos w$. It applies to all (e, M) with $0 \le e \le 1$ with one use of (8.16), which is a fourth-order extension of Newton's method. Mikkola also treated the hyperbolic case and a universal form of KE in a similar manner.

Most recently A. Nijenhuis, [Nijenhuis,1991], has described a scheme for the elliptic case which applies one iteration of a high-order extension of Newton's method to a one-step refinement of a rough starting value E_0 of E. A separate value E_0 is chosen for each of four subregions of the space $\{(e, M) : 0 \le e \le 1, 0 \le M \le \pi\}$; in the subregion $\{(e, M) : 0.55 \le e \le 1, 0 \le M \le 0.45\}$ the E_0 is the value used by Mikkola. While a step removed from the Schubert ideal, Nijenhuis' method can uniformly yield 16- or 19-place accuracy depending on the choice of order of the iteration method, with computational times as economical as any reported.

Most of the schemes we have described, and some we haven't, have been extensively tested, evaluated numerically, compared, and refined. The extent of the testing emphasizes very well how seriously modern astronautics and celestial mechanics regard the Kepler problem and its solution.

One description of a testing program, which embeds the Kepler problem in a most-modern setting, is the report [Klumpp,1991]. Also of note is the recent paper [Taff,1989] by L. Taff and T. Brennan.

In the latter paper, Taff and Brennan have used the approach of R. Serafin, [Serafin,1986], to provide a unifying theme to much of the derivation and testing of starting value-iteration combinations. To an accuracy of 10^{-9}, Taff and Brennan compared various starting values and iteration schemes by the average number of trigonometric evaluations required over (e, M)-space with $0 \leq e \leq 1$ to achieve this accuracy. Measuring efficiency from this viewpoint, they concluded that the most efficient scheme was to use starting value $E_0 = M + e$ with an iteration sequence proposed by J. Wegstein in 1968:

If $f(E) = E - e \sin E - M$, and $E_1 = f(E_0)$, let

$$E_{n+1} = E_n + \frac{E_n - E_{n-1}}{\left(\left[\dfrac{E_{n-1} - f(E_{n-1})}{E_n - f(E_n)} \right] - 1 \right)} - 1$$

$$n = 1, 2, 3, \ldots$$

It is hard to believe that there can be much room left for discovery and novelty in the subject of efficient numerical solutions of KE, but perhaps even now there is a new and clever experiment being devised to join this collection.

KE and Methods for Transcendental Equations

Mathematical models for many problems of applied science lead to transcendental equations—equations whose solutions cannot be described by polynomial equations. The eigenvalue equations of partial differential operators, for example, are often transcendental. In contrast to numerical methods like iteration which may be used to solve such equations approximately and repeatedly, exact methods which can express solutions in analytical form offer the chance to describe how the solutions vary as parameters in the equations change. KE, as one of the earliest transcendental equations to be studied, earns a preeminent place among transcendental equations and serves as an important test case for any methods which treat them.

Of the variety of methods available, we will treat only four:

1. The method of Burniston and Siewert

2. The method of Ioakimidis and Papadakis

3. The method of Delves-Lyness

4. Newton's method for power series

The first two have been used in the literature to tackle KE, [Siewert,1972] and [Ioakimidis,1985]. The third is an attractive but unstable method for solving equations which has been tried only recently on KE, [Bina,1990]. The results of the fourth applied to KE we believe to be new. Since careful description of the methods makes mathematical demands ranging from moderate to severe, we have made what we hope is a sensible choice of how much detail to include here and how much to put back in the Appendices with fuller references.

1. The Burniston-Siewert Method

It was the insight of E. Burniston and C. Siewert to consider solving transcendental equations by recasting them in the form of Riemann boundary value problems. The latter, also associated with the boundary value problems of Hilbert and Privaloff, have been studied in a variety of contexts. The theory for the existence and character of canonical solutions for such problems has been well-developed since 1900, and their applications have been growing rapidly for the past 20 years. The original description of the Burniston-Siewert method appears in [Burniston,1973] and its application to KE in [Siewert,1972]. Although the mathematical sophistication required by the method is high, the analytical expression for solutions which result may allow one to:

- analyze how solutions depend on the underlying parameters;

- ascertain the number and location of solutions;

- provide refined approximate solutions which can be further polished by iterative methods.

The Burniston-Siewert method applied to KE is rich enough to give analytic expression to all solutions, real and complex, for all values of e. Here we'll describe only the real solution in the elliptical case, $0 < e < 1$; some description of the method and its workings will be left to Appendix K.

In the elliptical case, the form of the real solution changes as (e, M) is chosen from each of the three open regions

$$R_1 = \left\{ (e, M) : 0 < e < 1,\ e < \frac{\pi}{2} - M \right\}$$

$$R_2 = \left\{ (e, M) : 0 < e < 1,\ e > \frac{\pi}{2} - M \ \text{ and } \ e > M - \frac{3\pi}{2} \right\}$$

$$R_3 = \left\{ (e, M) : 0 < e < 1,\ e < M - \frac{3\pi}{2} \right\}$$

as drawn in Figure 8.1.

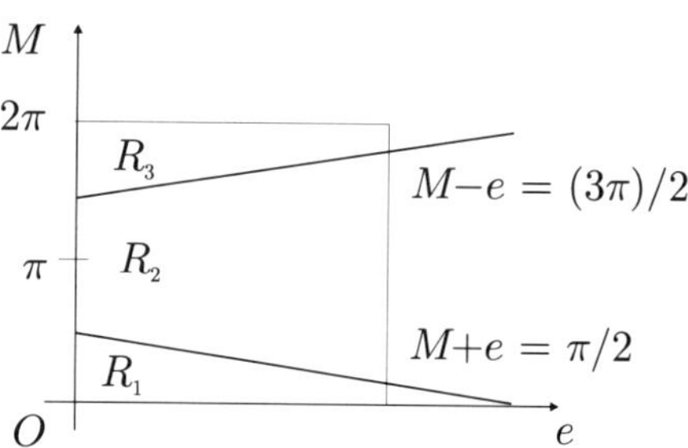

Figure 8.1

The solution for E is given by

$$E = M + e(2 - k)\left[S_{1k} + S_{2k} - \frac{1}{3}A_{2k}\right]^{-(1/2)}$$

when (e, M) is in R_k, $k = 1, 3$, and

$$E = M - e(M - \pi)\left[(e + 1)^2 - (M - \pi)^2\frac{2}{\pi}\int_0^1 F_2(t)dt\right]^{-(1/2)}$$

when (e, M) is in R_2.

The constants and functions in these expressions are defined as follows. For $k = 1, 2, 3$

$$F_k(t) = \arctan\left\{\frac{2(-1)^{k+1}tC(t)\left[e + (-1)^k\frac{\pi}{2}|t|\right]}{\left[e + (-1)^k\frac{\pi}{2}|t|\right]^2 - t^2[M - \pi\Delta(k)]^2 - t^2C(t)}\right\}$$

$$C(t) = \log\left[\sqrt{\frac{1}{t^2} - 1} + \frac{1}{|t|}\right]$$

$$\Delta(k) = k - 1$$

$$I_{jk} = \frac{1}{\pi}\int_0^1 t^j F_k(t)dt - \frac{2}{j + 1}$$

$$J(k) = -2(e - 1)^2\left[I_{1k}^2 + I_{3k}\right] + (e - 1)\left[\frac{2}{3}I_{1k} + \frac{3}{20}\right] - \frac{1}{36}$$

$$A_{2k} = 2I_{1k} - (e - 1)^2[M - \pi\Delta(k)]^{-2}$$

$$A_{1k} = 2I_{1k}^2 + I_{3k} + \left[-2(e-1)^2 I_{1k} + \frac{1}{3}(e-1)\right][M - \pi\Delta(k)]^{-2}$$

$$A_{0k} = \frac{4}{3}I_{1k}^3 + 4I_{1k}I_{3k} + 2I_{5k} + J(k)[M - \pi\Delta(k)]^{-2}$$

$$Q_k = \frac{1}{3}A_{1k} - \left(\frac{1}{3}A_{2k}\right)^2$$

$$D_k = \frac{1}{6}[A_{1k}A_{2k} - 3A_{0k}] - \left(\frac{1}{3}A_{2k}\right)^3$$

$$S_{jk} = \left[D_k - (-1)^j \left\{D_k^2 + Q_k^3\right\}^{(1/2)}\right]^{(1/3)}$$

This manner of displaying the solution of KE with all supporting details suppressed may do the result an injustice by hiding much of the information the method reveals about the number, location, and dependence of the solutions on e and M. Burniston and Siewert did verify that this solution does produce accurate numerical results, and their method has attracted attention for its successes with a variety of transcendental equations, [Henrici,1986, pp. 183–193].

2. The Ioakimidis-Papadakis Method

N. Ioakimidis thought to obtain the solution of an equation $f(x) = 0$ in analytical form by embedding the equation in a system

$$\left\{ \begin{array}{l} F(x,y) = f(x) = 0 \\ G(x,y) = y = 0 \end{array} \right\}$$

applying an old method of Picard, and using integration by parts on the result. Ioakimidis and K. Papadakis, [Ioakimidis,1985], produced further expressions by more integrations by parts and applied them to KE in the form

$$f(x) = x - e\sin(x + M) = 0 \tag{8.17}$$

where $x = E - M$.

With the notation

$$g(x)\big|_a^b = g(b) - g(a)$$

for an arbitrary positive number p they used their method on (8.17) to find

$$x = \pm \frac{1}{\pi}\left\{ \int_{-e}^{e} \left[\frac{pt f(t) f''(t) - (f'(t))^2}{(f(t))^2 + p^2 (f'(t))^2} \right.\right.$$

$$\left. + \arctan\left(\frac{pf'(t)}{f(t)}\right)\right] dt \tag{8.18}$$

$$\left. - \left[t \arctan\left(\frac{pf'(t)}{f(t)}\right)\right]\Big|_{-e}^{e} \right\}$$

and $E = x + M$.

Testing (8.18) by Gaussian quadrature won't give accurate results unless a high number of nodes is used, but higher-order variations of (8.18) arising from several more integrations by parts were more successful.

3. The Delves-Lyness Method

The method of Delves and Lyness was described in [Delves,1967] to solve equations $f(z) = 0$ when f is analytic. For those with a modest background in complex analysis, the method is appealing and reasonably simple to describe. However, as a practical method for solving equations it simply doesn't work very well, and efforts to improve its effectiveness have not had much success.

If Γ is a simple closed curve in $\mathcal{C}$ which passes through no zero of analytic function f, and if the zeros of f inside Γ are known to be

$$z_1, z_2, \ldots, z_N$$

the Cauchy residue theory indicates that

$$\frac{1}{2\pi i} \oint_{\Gamma} \frac{f'(z)}{f(z)} dz = N \tag{8.19}$$

and

$$\frac{1}{2\pi i} \oint_{\Gamma} \frac{z^k f'(z)}{f(z)} dz = z_1^k + z_2^k + \cdots + z_N^k \tag{8.20}$$

$$k = 1, 2, 3, \ldots, N.$$

If we let $s_k = z_1^k + z_2^k + \cdots + z_N^k$, $k = 1, 2, 3, \ldots, N$, and the quantities $\sigma_1, \sigma_2, \ldots, \sigma_N$ are defined by

$$\left\{ \begin{aligned} p(z) &= z^N - \sum_{k=1}^{N} \sigma_k z^{N-k} \\[2mm] &= (z - z_1)(z - z_2)\ldots(z - z_N) \end{aligned} \right\} \tag{8.21}$$

then the numbers $\{s_k\}_1^N$ and $\{\sigma_k\}_1^N$ are related by *Newton's identities*

$$\left\{ \begin{array}{c} s_1 + \sigma_1 = 0 \\[2ex] s_k + \sum_{j=1}^{k-1} s_{k-j}\sigma_j + k\sigma_k = 0 \\[2ex] k = 2, 3, \ldots, N. \end{array} \right\} \tag{8.22}$$

Delves and Lyness thought to compute N in (8.19) for a given Γ and f, then to compute $\{s_k\}_1^N$ in (8.20), and then to use (8.22) to compute $\{\sigma_k\}_1^N$. The polynomial (8.21) thereby determined would have as its zeros all the solutions of $f(z) = 0$ inside Γ and these could be computed by any effective polynomial solver.

Of course, since it is known that errors in the computed zeros of a polynomial can be substantial if there are even small errors in its coefficients, it is to be expected that the Delves-Lyness method would be unstable. Errors in $\{\sigma_k\}_1^N$ coming from errors in the computation of $\{s_k\}_1^N$ arise not only from errors in numerical integration but also when f has one or more zeros close to Γ, whether inside Γ or not.

Recently S. Bina, who was making a number of numerical experiments with the Delves-Lyness method, included the solving of KE among them, [Bina,1990]. He found in the elliptic case, even with $N = 1$ in (8.20), that the solutions of KE in this manner can be quite inaccurate unless the curve Γ is carefully chosen—circles of small radius with center already near the solution worked best!

4. Newton's Method for Power Series

We'll conclude our look at modern treatments of KE, as well as our entire study, on a retrospective note by returning to Newton's method. That what is so old may yet have a modern aspect is satisfying and a bit surprising.

Newton's method, as we mean it here, is an algorithm for formal power series which Newton himself used in about 1669, [Whiteside,1968, pp. 206–247]. Presently it is a part of standard technique in developing fast algorithms to evaluate special functions in scientific computing, [Henrici,1986, pp. 72–80].

Formal power series is an algorithmic treatment of manipulative aspects of power series with no immediate concern for matters of convergence. With each sequence $\{a_n\}_0^\infty$ of complex numbers we associate a formal power series (fps)

$$f(x) = \sum_{n=0}^{\infty} a_n x^n$$

(in which the variable x need have no significance). If $\mathfrak{S}$ is the collection of fps, and if

$$f(x) = \sum_{n=0}^{\infty} a_n x^n, \quad g(x) = \sum_{n=0}^{\infty} b_n x^n$$

are chosen from $\mathfrak{S}$, we can define

$$(f+g)(x) = \sum_{n=0}^{\infty}(a_n + b_n)x^n$$

$$(fg)(x) = \sum_{n=0}^{\infty} c_n x^n \quad \text{where} \quad c_n = \sum_{k=0}^{n} a_k b_{n-k}.$$

The elements of $\mathfrak{S}$ corresponding to sequences $\{a_n\}_0^\infty$ with $a_0 = 0$, $a_1 \neq 0$ are called *almost units,* and we let $\wp$ be the subcollection of $\mathfrak{S}$ made up of all the almost units. In $\wp$ the composition operator "$\circ$" is well-defined, and in fact $\wp$ is a group with respect to composition with identity $f(x) = x$. We denote by $f^{[-1]}$ the inverse of f in $\wp$ so that

$$(f^{[-1]} \circ f)(x) = x = (f \circ f^{[-1]})(x).$$

If f and g lie in $\wp$, the equation

$$(f \circ w)(x) = g(x) \tag{8.23}$$

has a unique solution in $\wp$ given by

$$w(x) = (f^{[-1]} \circ g)(x). \tag{8.24}$$

To describe Newton's method requires some additional notation. With

$$f(x) = \sum_{n=0}^{\infty} a_n x^n$$

we let

$$f'(x) = \sum_{n=0}^{\infty}(n+1)a_n x^n \quad \text{and} \quad (f)_k(x) = \sum_{n=0}^{k} a_n x^n.$$

Thus, *Newton's Method for fps* is stated as follows: For $m = 0, 1, 2, \ldots, w(x)$ in (8.24) satisfies

$$(w)_{2m+1} = (w)_{2m} - \left[\frac{(f \circ (w)_{2m}) - g}{f' \circ (w)_{2m}} \right]_{2m+1}. \tag{8.25}$$

A fuller description of Newton's method is given in Appendix L.

Realizing that we had encountered no report that (8.25) had been applied to KE, we undertook to do so. The results are informative.

1. When $x = E - M$ and KE is rewritten

$$\frac{x}{\sin(M + x)} = e \qquad (8.26)$$

and M is not an integral multiple of π, we may use

$$f(x) = \frac{x}{\sin(M + x)} \quad \text{and} \quad g(x) = x$$

from $\wp$ to write (8.26) as

$$f \circ w(e) = e \qquad (8.27)$$

and apply Newton's method (8.25) to obtain an fps for

$$w(e) = M + \sum_{n=1}^{\infty} a_n(M) e^n. \qquad (8.28)$$

The coefficients in this case are identical with those of Lagrange's solution of KE

$$a_n(M) = \frac{1}{n!} \frac{d^{n-1}}{dM^{n-1}} (\sin^n M), \quad n = 1, 2, 3, \dots .$$

Explicit formulas for $a_1(M)$, $a_2(M)$, $\dots$, $a_{10}(M)$ are listed in Appendix D.

2. Since

$$z = h(e) = \frac{e \exp(\sqrt{1 - e^2})}{1 + \sqrt{1 - e^2}}$$

lies in $\wp$, so does $e = h^{[-1]}(z)$ and (8.26) may be viewed in the form (8.23) with

$$f(x) = \frac{x}{\sin(M + x)} \quad \text{and} \quad g(x) = h^{[-1]}(x)$$

so that solution by (8.25) produces

$$w(z) = M + \sum_{n=1}^{\infty} b_n(M) z^n. \qquad (8.29)$$

Of course, z is the Levi-Civita variable and the coefficients $\{b_n(M)\}$ exactly match those of the Levi-Civita solution of KE. In Appendix E explicit forms of $b_1(M)$, $b_2(M)$, $\dots$, $b_{10}(M)$ appear.

3. With $f(x) = x - e\sin x$ and $g(x) = x$ in $\wp$, KE may be written as

$$(f \circ E)(M) = M.$$

Application of (8.25) here gives the fps solution

$$E(M) = \sum_{n=1}^{\infty} c_n(e) M^n$$

and actual calculations determined that these coefficients

$$\{c_n(e)\}$$

are those of the Lie-series solution of Stumpff:

$$c_n(e) = \frac{1}{n!} D^n x\big|_{x=0}$$

where

$$D^1 x = \frac{1}{1 - e\cos x}, \quad D^{n+1} x = \left(\frac{1}{1 - e\cos x}\right) \frac{d}{dx}(D^n x).$$

These coefficients for $n = 1, 2, \ldots, 10$ are displayed in Appendix G. So it happens that with the use of Newton's method of 1669 we are able to display a solution of KE which was not fully described until 1968.

It is worth a few words to explain why the well-known solution (3.8) in terms of Bessel functions is not accessible through (8.25).

Although we may write (3.8) in the form of an infinite series in powers of $w = \exp(iM)$ by substituting

$$\sin nM = \frac{1}{2i}(w^n - w^{-n})$$

and rearranging, the resulting series is not in $\Im$.

Cauchy, in [Cauchy,1841], did rewrite (3.8) as

$$E = M + \sum_{\substack{n=-\infty \\ n \neq 0}}^{\infty} A_n(e) \exp(inM)$$

with

$$A_n(e) = \frac{e}{2\pi} \int_0^{2\pi} [\exp(-inM)\sin M$$

$$(1 - e\cos M)\exp(ien\sin M)]dM.$$

One can use residues to show that, in fact,

$$A_n(e) = \frac{-iJ_n(ne)}{n}.$$

There is an operational path leading directly from Lagrange's series (8.28) to (3.8); it was given by O. Fleckenstein, [Fleckenstein,1941], and it completed Lagrange's original attempt to rearrange (8.28) into (3.8) which we described in Chapter 3.

Appendix A

Geometric Parallax and KE

If observers at different positions on the earth's surface wish to identify the location of an astronomical object by a set of coordinates independent of their own positions which will allow a third observer to locate the same object, coordinates relative to the earth's center (*geocentric coordinates*) may be used.

The geocentric parallax of an object relative to an observer is the angle measured from the object between the lines drawn from the object to an observer on the earth's surface and to the center of the earth. For objects further from the earth than the moon, the geocentric parallax is usually less than $1°$.

For a given object, let α be the right ascension, δ the declination, r the distance, and ϕ the astronomical latitude relative to the earth's center, and let α', δ', r', ϕ' be the corresponding measurements relative to an observer on the earth's surface. We presume α', δ', ϕ', ϕ are observables, as are ρ, the distance between observer and earth's center, and θ, the observer's local sidereal time.

The specific problem we address is to determine α and δ from the observed quantities. This problem is sometimes called *the parallax problem* although we are not calculating the geocentric parallax.

If angle γ is defined by

$$\tan \gamma = \tan \phi \sec(\alpha' - \theta)$$

γ is also an observable, and using analytic geometry in spherical coordinates one finds, [Roy,1988, pp. 64–68],

$$\alpha' - \alpha = \frac{\rho}{r'} \cos \phi \sec \delta' \sin(\alpha' - \theta) \tag{A.1}$$

$$\delta' - \delta = \frac{\rho}{r'} \sin \phi \csc \gamma \sin(\delta' - \gamma). \tag{A.2}$$

In the special case where the observer is on the equator and the object is viewed at the observer's local horizon, $\phi = 0$ and the geocentric parallax is called the *equatorial horizontal parallax*. In this case $\delta' = \delta$ and equation (A.1) can be rewritten

$$\alpha - \theta = (\alpha' - \theta) - \left[\frac{\rho}{r'}\sec\delta\right]\sin(\alpha' - \theta). \qquad (A.3)$$

Since ρ/r' will be quite small, the relation between α and α' is governed by Kepler's equation.

Appendix B

Error in the Horrocks Solution

If E_H is the Horrocks solution and M is the mean anomaly, we repeat equation (2.4):

$$\tan\left(E_H - \frac{M}{2}\right) = \left(\frac{1+e}{1-e}\right)\tan\frac{M}{2}. \tag{2.4}$$

With

$$k = \frac{1+e}{1-e},$$

if $\tan y = k\tan x$, then

$$\sec^2 y \frac{dy}{dx} = k\sec^2 x$$

$$\frac{dy}{dx} = \frac{k\sec^2 x}{1+\tan^2 y} = \frac{k\sec^2 x}{1+k^2\tan^2 x}$$

and y satisfies the initial value problem

$$\frac{dy}{dx} = \frac{k}{\cos^2 x + k^2\sin^2 x} \tag{B.1}$$

$$y(0) = 0.$$

Let $z = \exp(ix)$ so that

$$\frac{k}{\cos^2 x + k^2 \sin^2 x} = \frac{4kz^2}{(z^2 + 1)^2 - k^2(z^2 - 1)^2}$$

$$= \left(\frac{4kz^2}{1 - k^2}\right) \left[\frac{1}{\left(z^2 - \dfrac{k+1}{k-1}\right)\left(z^2 - \dfrac{k-1}{k+1}\right)} \right]$$

$$= \frac{1}{1 - \dfrac{e}{z^2}} + \frac{ez^2}{1 - ez^2}$$

where

$$\frac{k-1}{k+1} = e.$$

Then

$$\frac{k}{\cos^2 x + k^2 \sin^2 x} = 1 + \sum_{n=1}^{\infty} e^n \left(z^{2n} + z^{-2n}\right)$$

$$= 1 + 2\sum_{n=1}^{\infty} e^n \cos 2nx.$$

Consequently (B.1) has the solution

$$y = x + \sum_{n=1}^{\infty} \frac{e^n}{n} \sin 2nx. \tag{B.2}$$

In (2.4) with

$$y = E_H - \frac{M}{2} \quad \text{and} \quad x = \frac{M}{2},$$

$$E_H = M + \sum_{n=1}^{\infty} \frac{e^n}{n} \sin nM. \tag{B.3}$$

Through order 3 in e, Lagrange's solution of KE is

$$E \approx M + e \sin M + \frac{1}{2}e^2 \sin 2M + e^3 \left(\frac{3}{8} \sin 3M - \frac{1}{8} \sin M\right) \tag{B.4}$$

so we can determine

$$E_H - E \approx e^3 \left(\frac{1}{3} \sin 3M - \frac{3}{8} \sin 3M + \frac{1}{8} \sin M \right)$$

$$\approx e^3 \left(\sin M - \frac{4}{3} \sin^3 M - \frac{9}{8} \sin M \right.$$

$$\left. + \frac{3}{2} \sin^3 M + \frac{1}{8} \sin M \right)$$

$$\approx \frac{1}{6} e^3 \sin^3 M$$

which is the estimate (2.5).

Appendix C

Machin's E_1

To account for Machin's initial choice, E_0 of E, given e and M, we follow [Brinkley,1803, pp. 109–114].

In KE let $E = nA$ so that

$$M = nA - e\sin nA.$$

With $A = \arcsin s$ rewrite KE in powers of s:

$$\left\{ \begin{aligned} M &= n\left[s + \frac{s^3}{3!} + \frac{3^2 s^5}{5!} + \frac{3^2 5^2 s^7}{7!} + \cdots\right] \\[2mm] &\quad -en\left[s + \frac{(1-n^2)s^3}{3!}\right. \\[2mm] &\quad \left. + \frac{(1-n^2)(3^2-n^2)s^5}{5!} + \cdots\right] \end{aligned} \right\} \tag{C.1}$$

Machin noticed that the s^5-terms in (C.1) are removed if $e(n^2-1)(n^2-3^2) = 9$, or

$$n = \sqrt{5 + \sqrt{16 + \frac{9}{e}}}. \tag{C.2}$$

Through terms of order 5 in s, with this choice of n (C.1) becomes

$$M = n\left[(1-e)s + \frac{e(n^2-1)+1}{3!}s^3\right]. \tag{C.3}$$

Machin proposed to solve (C.3) for s and take $E_0 = n\arcsin s$ with n as in (C.2).

It was Machin's plan then to correct E_0 to

$$E_1 = E_0 - \left(\frac{E_0 - e\sin E_0 - M}{1 - e\cos E_0} \right)$$

which is one iteration by Newton's method.

To indulge our curiosity about Machin's E_1, we made the following calculations. For $M = 0.05\pi,\ 0.1\pi,\ \ldots,\ 0.5\pi$ and the values of e in the table elow we

- solved (C.3) for s;

- calculated E_0, E_1, $E_1 - e\sin E_1 - M$;

- set $A(e)$ equal to the average value of $E_1 - e\sin E_1 - M$ over the ten M-values for each e.

e	$A(e)$
0.1	4.56×10^{-7}
0.2	8.81×10^{-6}
0.3	5.46×10^{-5}
0.4	2.09×10^{-4}
0.5	6.00×10^{-4}
0.6	1.43×10^{-3}
0.7	2.95×10^{-3}
0.8	5.48×10^{-3}
0.9	9.39×10^{-3}
0.92	0.0104
0.94	0.0115
0.96	0.0138
0.98	0.0142
0.995	0.0154

In numerical experiments reported in Chapter 8, typical starting values tested with Newton's method are

$$E_0 = M,\ E_0 = M + e,\ E_0 = M + \frac{e\sin M}{1 + \sin M - \sin(M + e)}$$

all of which are simpler to obtain than Machin's E_1. However, with each of them the corresponding values of $A(e)$ as described above are considerably inferior to those in this table. So Machin's E_1 shows itself to have been clever and quite effective.

Appendix D

Coefficients for the Lagrange Solution

In the Lagrange solution

$$E - M = \sum_{n=1}^{\infty} a_n(M) e^n$$

where

$$a_n(M) = \frac{1}{n!} \frac{d^{n-1}}{dM^{n-1}} (\sin^n M).$$

The coefficients $\{a_n(M)\}$ may be obtained by Newton's method (Appendix L) or by repeated differentiation. In the table below, obtained with the software package MATHEMATICA, Wolfram Research, Inc., Enhanced Version 1.2, 1988, we have set $S = \sin M$ and $C = \cos M$.

n	$a_n(M)$
1	S
2	CS
3	$C^2 S - \frac{1}{2} S^3$
4	$C^3 S - \frac{5}{3} C S^3$
5	$C^4 S - \frac{11}{3} C^2 S^3 + \frac{13}{24} S^5$
6	$C^5 S - \frac{20}{3} C^3 S^3 + \frac{47}{15} S^5$

$$7 \qquad C^6 S - \frac{65}{6} C^4 S^3 + \frac{1291}{120} C^2 S^5 - \frac{541}{720} S^7$$

$$8 \qquad C^7 S - \frac{49}{3} C^5 S^3 + \frac{427}{15} C^3 S^5 - \frac{1957}{315} S^7$$

$$9 \qquad C^8 S - \frac{70}{3} C^6 S^3 + \frac{1281}{20} C^4 S^5 - \frac{36619}{1260} C^2 S^7 + \frac{9509}{8064} S^9$$

$$10 \qquad C^9 S - 32 C^7 S^3 + \frac{644}{5} C^5 S^5 - \frac{6368}{63} C^3 S^7 + \frac{5141}{405} S^9.$$

In 1820 F.T. Schubert would have paid a great price to have had such software.

Appendix E

Coefficients of the Levi-Civita Solution

In the Levi-Civita solution

$$E = \sum_{n=0}^{\infty} L_n(M) z^n \qquad (E.1)$$

where

$$z = \frac{e \exp(\sqrt{1 - e^2})}{1 + \sqrt{1 - e^2}}.$$

To obtain explicit formulas for the coefficients $\{L_n(M)\}$ as functions of M, one may use Newton's method (Appendix L) or the scheme we describe below; in either case the computational cost can become high. We'll derive a recursive scheme and use it to determine $L_0(M)$, $L_1(M)$, ..., $L_{10}(M)$.

To be able to write E in powers of z we must first be able to write

$$e = \sum_{k=0}^{\infty} b_k z^k \qquad (E.2)$$

where $0 \le e < 1$. With standard results about inversion of power series, and noting that $b_{2k} = 0$ for $k = 0, 1, 2, \ldots$, we obtained

$$\left\{ b_{2k+1} = (-1)^k \left(\tfrac{2}{2k+1} \right) \exp\left(-(2k+1)\right) \right.$$

$$\left. \cdot \sum_{j=0}^{k} \left[\frac{(-k)_j (2k+1-j) \{2(2k+1)\}^j}{j!(j+1)!} \right] \right\} \qquad (E.3)$$

where

$$(-k)_j = (-k)(-k+1) \cdot \cdots \cdot (-k+j-1)$$

$$k = 0, 1, 2, \ldots$$

With (E.1) and (E.2), KE reads

$$M = \sum a_k z^k - \left(\sum b_k e^k \right)$$

$$\cdot \left[\frac{\exp\left(\sum (ia_k) z^k\right) - \exp\left(\sum (-ia_k) z^k\right)}{2i} \right] \tag{E.4}$$

where "$\sum$" denotes "$\sum\limits_{k=0}^{\infty}$" and we let $a_k = L_k(M)$ temporarily.

If

$$\sum c_k z^k = \exp\left[\sum (ia_k) z^k\right]$$

and

$$\sum C_k z^k = \exp\left[\sum (-ia_k) z^k\right]$$

one can verify that

$$c_0 = \exp(ia_0), C_0 = \exp(-ia_0)$$

$$c_{k+1} = i \sum_{j=0}^{k} \left(1 - \frac{j}{k+1}\right) a_{k+1-j} c_j$$

$$C_{k+1} = -i \sum_{j=0}^{k} \left(1 - \frac{j}{k+1}\right) a_{k+1-j} C_j$$

$$k = 0, 1, 2, \ldots$$

With

$$d_k = \frac{c_k - C_k}{2i}$$

(E.4) becomes

$$\sum a_k z^k - \left(\sum b_k z^k\right)\left(\sum d_k z^k\right) = M$$

and

$$\sum a_k z^k - \sum_{k=0}^{\infty} \left\{ \left(\sum_{j=0}^{\infty} b_j d_{k-j}\right) z^k \right\} = M.$$

Thus

$$\left\{ \begin{array}{l} a_0 = M, \quad a_k = \sum_{j=0}^{\infty} b_j d_{j-k} \\[2em] k = 1, 2, 3, \ldots \end{array} \right\} \tag{E.5}$$

By direct calculation from (E.3)

$$b_1 = 0.7357589$$
$$b_3 = 0.0995741$$
$$b_5 = 0.0404277$$
$$b_7 = 0.0224931$$
$$b_9 = 0.0145624$$

In calculating the coefficients a_j, b_j, c_j, C_j, d_j for $j = 0, 1, 2, \ldots, k-1$, which are needed to determine a_k, we used the software package MACSYMA, Symbolics, Inc., Version 11, 1985. The results for

$$\{a_k = L_k(M) : k = 0, 1, \ldots, 10\}$$

are listed below with each numerical factor rounded to six decimals, and the abbreviations $S = \sin M$ and $C = \cos M$.

$$L_0(M) = M$$
$$L_1(M) = 0.735759 S$$
$$L_2(M) = 0.541341 SC$$
$$L_3(M) = -0.199148 S^3 + 0.398297 SC^2 + 0.099574 S$$
$$L_4(M) = 0.488417 S^3 C + 0.29305 SC^3 + 0.146525 SC$$
$$L_5(M) = 0.116791 S^5 - 0.790586 S^3 C^2 - 0.080855 S^3$$
$$\qquad\qquad +0.215614 SC^4 + 0.161711 SC^2 + 0.040428 S$$
$$L_6(M) = 0.497072 S^5 C - 1.057601 S^3 C^3 - 0.2644 S^3 C$$
$$\qquad\qquad +0.15864 SC^5 + 0.15864 SC^3 + 0.069405 SC$$
$$L_7(M) = -0.087703 S^7 + 1.255722 S^5 C^2 + 0.07903 S^5$$
$$\qquad\qquad -1.264476 S^3 C^4 - 0.534971 S^3 C^2 - 0.04377 S^3$$
$$\qquad\qquad +0.116721 SC^6 + 0.145901 SC^4 + 0.087541 SC^2$$
$$\qquad\qquad +0.022493 S$$
$$L_8(M) = -0.533537 S^7 C + 2.444673 S^5 C^3 + 0.403629 S^5 C$$
$$\qquad\qquad -1.402681 S^3 C^5 - 0.858784 S^3 C^3 - 0.161022 S^3 C$$

$$+0.085878SC^7 + 0.128818SC^5 + 0.096613SC^3$$
$$+0.04115SC$$
$$L_9(M) = 0.074508S^9 - 1.83635S^7C^2 - 0.083085S^7$$
$$+4.047052S^5C^4 + 1.189605S^5C^2 + 0.053478S^5$$
$$-1.474336S^3C^6 - 1.197898S^3C^4 - 0.362002S^3C^2$$
$$-0.027644S^3 + 0.063186SC^8 + 0.110575SC^6$$
$$+0.098728SC^4 + 0.055288SC^2 + 0.014562S$$
$$L_{10}(M) = 0.59013S^9C - 4.699132S^7C^3 - 0.577651S^7C$$
$$+5.987851S^5C^5 + 2.646804S^5C^3 + 0.300439S^5C$$
$$-1.487665S^3C^7 - 1.518658S^3C^5 - 0.639231S^3C^3$$
$$-0.108153S^3C + 0.04649SC^9 + 0.092979SC^7$$
$$+0.095885SC^5 + 0.064892SC^3 + 0.027543SC$$

Since to our knowledge no numerical test has been made of the Levi-Civita solution, we have made a simple set of calculations to give some idea of its results.

With e chosen as in the table below we have calculated for each

$$M_j = \frac{\pi j}{50}, \quad j = 1, 2, 3, \ldots, 50$$

the quantities

$$E^* = \sum_{k=0}^{10} L_k(M)z^k$$

and E_N, the solution by Newton's method iterated to the point where successive iterations differed by less than 10^{-10}.

The quantity

$$B(e) = \sum_{j=1}^{50} \frac{100|E^* - E_N|}{E_N}$$

summed over the 50 M-values represents average percentage absolute relative error of E^* with respect to E_N for each e.

e	$B(e)$
0.1	2.12×10^{-6}
0.18	1.05×10^{-5}
0.26	5.29×10^{-5}
0.34	4.0×10^{-4}
0.42	3.29×10^{-3}
0.5	0.018
0.58	0.073
0.66	0.225
0.74	0.554
0.82	1.13
0.9	2.0
0.98	3.36

While the comparison is crude and tends to understate the actual errors $|E^* - E_N|$ for individual (e, M)-pairs, it does show that the average relative error in E^* increases with e. While the series (E.1) converges for values of z up to 1, when e exceeds 0.66274 (the radius of convergence of Lagrange's solution), z exceeds 0.8 and the rate of convergence of the Levi-Civita solution falters.

Appendix F

Autonomous Differential Equations and Lie-Series

There are several elementary techniques known to solve first-order ordinary differential equations of the form

$$\frac{dx}{dt} = f(x, t).$$

A special simplicity arises when the differential equation is *autonomous*—that is, has the form

$$\frac{dx}{dt} = f(x)$$

where the independent variable t does not appear explicitly in the equation. Then, at least formally, the variables separate

$$\frac{dx}{f(x)} = dt$$

and one can hope to recover solutions by direct integration.

Let A be an $n \times n$ matrix of constants with n distinct eigenvalues λ_1, λ_2, ..., λ_n. If C_1, C_2, ..., C_n are the eigenvectors of A corresponding to λ_1, λ_2, ..., λ_n respectively, the matrix C with columns C_1, C_2, ..., C_n is nonsingular, and it is known that

$$A = C \begin{pmatrix} \lambda_1 & 0 & \dots & 0 \\ 0 & \lambda_2 & \dots & 0 \\ & & \dots & \\ 0 & 0 & \dots & \lambda_n \end{pmatrix} C^{-1}$$

and the *exponential matrix* of A, $\exp(tA)$, will be

$$\exp(tA) = \sum_{n=0}^{\infty} \frac{t^n}{n!} A^n$$

$$= C \begin{pmatrix} \exp(\lambda_1 t) & 0 & \dots & 0 \\ 0 & \exp(\lambda_2 t) & \dots & 0 \\ & & \dots & \\ 0 & 0 & \dots & \exp(\lambda_n t) \end{pmatrix} C^{-1}$$

If $X(t) = (x_1(t), x_2(t), \dots, x_n(t))$ is an n-vector of unknown functions which satisfy collectively the autonomous system

$$X'(t) = AX(t)$$

with initial condition $X(0) = B$, then there is a unique solution for $X(t)$ given by

$$X(t) = \exp(tA)B = \sum_{n=0}^{\infty} \frac{t^n}{n!} (A^n B). \tag{F.1}$$

Equation (F.1) provides a simple context for the definition (3.22) of the Lie-series of an operator D on a space of functions $\Im$: in this case, operator D is multiplication by matrix A, $\Im$ is the collection of n-vector functions of a single variable, and the Lie-series is the matrix product $\exp(tA)B$.

In the case of KE, where $M = E - e \sin E$, there is an associated autonomous differential equation

$$\frac{dE}{dM} = 1 - e \cos E. \tag{F.2}$$

With initial condition $E(0) = 0$ and operator

$$D(x, f) = \left(\frac{1}{1 - e \cos x} \right) \frac{d}{dx}(f)$$

equation (F.2) will have the solution

$$L(D, 0, M) = \exp(MD)(0) = \sum_{n=0}^{\infty} \frac{M^n}{n!} D^n(0)$$

which is Stumpff's solution (3.23).

Appendix G

Coefficients of the Lie-Series Solution

The Lie-series solution of KE described in Chapter 3 has the form

$$E = \sum_{n=0}^{\infty} \frac{S^n M^n}{n!} \tag{G.1}$$

where S^n denotes the n-th iterate of the operator

$$D(\ ) = \left\{ \left(\frac{1}{1 - e \cos x} \right) \frac{d}{dx}(\ ) \right\} \Big|_{x=0}.$$

One can show that $S^{2k} = 0$, $k = 0, 1, 2, \ldots$, and we used MATHEMAT-ICA, Wolfram Research, Inc., Enhanced Version 1.2, 1988, to calculate S^{2k+1} for $k = 0, 1, \ldots, 7$.

$$S^1 = \frac{1}{1 - e}$$

$$S^3 = \frac{-e}{(1 - e)^4}$$

$$S^5 = \frac{e(1 + 9e)}{(1 - e)^7}$$

$$S^7 = \frac{-e(1 + 54e + 225e^2)}{(1 - e)^{10}}$$

$$S^9 = \frac{e(1 + 243e + 4131e^2 + 11025e^3)}{(1 - e)^{13}}$$

$$S^{11} = \frac{-e(1 + 1008e + 50166e^2 + 457200e^3 + 893025e^4)}{(1-e)^{16}}$$

$$S^{13} = \frac{e\left[\begin{array}{l} 1+\ 4077e + 520218e^2 + 11708154e^3 \\[4pt] +\ 70301925e^4 + 108056025e^5 \end{array}\right]}{(1-e)^{19}}$$

$$S^{15} = \frac{-e\left[\begin{array}{l} 1+\ 16362e + 5020623e^2 + 243313164e^3 \\[4pt] +\ 3274844175e^4 + 14427513450e^5 \\[4pt] +\ 18261468225e^6 \end{array}\right]}{(1-e)^{22}}.$$

The solution (G.1) presumes the initial condition $E(0) = 0$. From the form of the calculated $S_k's$, it is feasible only for small e.

To display some numerical results comparing

$$E^* = \sum_{k=0}^{15} \frac{S^k M^k}{k!}$$

with E_N, the approximate solution by Newton's method iterated for each (e, M)-pair until successive iterations differ by less than 10^{-10}, for each e-value in the table below we calculated

$$C(e) = \max\left\{ M : |E^*(e, M) - E_N(e, M)| < 1.0x10^{-5} \right\}$$

e	$C(e)$	e	$C(e)$
0.05	1.69646	0.55	0.2513
0.1	1.2566	0.6	0.1885
0.15	1.0053	0.65	0.1257
0.2	0.8168	0.7	0.1257
0.25	0.6912	0.75	0.0628
0.3	0.5655	0.8	0.0628
0.35	0.5027	0.85	0.0157
0.4	0.4398	0.9	0.0157
0.45	0.3142	0.95	0.0079
0.5	0.2513		

Appendix H

Binary Systems and KE

Suppose objects A and B comprise a binary system. Assuming A is fixed and B moves about A, we wish to calculate the apparent position of B relative to A in a plane $\mathfrak{R}$ with origin at A and perpendicular to an observer's line of sight to A. We'll measure the position of B by polar coordinates (ρ, θ) where θ is measured in the counterclockwise direction from the upper half line in $\mathfrak{R}$ through A perpendicular to the observer's line of sight.

It is assumed that the orbit of B about A is an ellipse with focus at A lying in a plane $\mathfrak{I}$ making angle i with $\mathfrak{R}$. On a sphere centered at A the planes $\mathfrak{R}$ and $\mathfrak{I}$ intersect in circles R and F. Relative to the direction of motion of B about A, we denote the ascending node of B by the point L in the orbit of B which lies on the intersection of R and F at which the position of B changes from below plane $\mathfrak{R}$ to above plane $\mathfrak{R}$. If P denotes the point in its orbit when B is closest to A, let ω be the directed angle from line AL to line AP, and let Ω be the value of θ determined by L.

The position of B in its orbit is measured by polar coordinates (r, v) relative to A, where v is the true anomaly calibrated so that $v = 0$ when B is at P.

If the quantities a, e, M, E have their usual meanings for this orbit, where $E = M = 0$ at P, then

$$M = E - e \sin E$$

$$\tan \frac{v}{2} = \sqrt{\frac{1 + e}{1 - e}} \tan \frac{E}{2}$$

$$r = a(1 - \cos E).$$

From e and M, r and v may be determined as usual.

Corresponding to each position (r, v) in the orbit, the coordinates (ρ, θ) in plane $\Re$ are calculated by

$$\tan(\theta - \Omega) = \tan(v + \omega)\cos i$$

$$\rho = r\cos(v + \omega)\sec(\theta - \Omega).$$

Appendix I

Hyperbolas and Battin's Universal KE

Although Kepler's second law, the "equal area" law, was described for elliptical orbits about the sun, Newton showed that it holds for any object in orbit about the sun subject to a central force of attraction and must include the parabolic and hyperbolic cases. The argument is neatly described in [Pollard,1976, pp. 2–5].

For an object in orbit about the sun with position $R(t)$ and velocity $V(t)$ at time t, $R(t) \times V(t) = C$ is a constant vector. With

$$c = |C|, \quad r(t) = |R(t)| \quad \text{and} \quad v(t) = |V(t)|$$

$$r(t)v(t) = c$$

and the law of conservation of energy implies

$$r^2 \left(\frac{dr}{dt} \right)^2 + c^2 = 2(\mu r + h r^2) \tag{I.1}$$

where μ and h are constants.

In trying to determine $R(t)$, the essential role of the eccentric anomaly and KE is to establish an artificial "time" E in terms of which position is more conveniently expressed, and then to determine t from E. Choosing $E = E(t)$ so that

$$r(t)\frac{dE}{dt} = k,$$

a constant, and

$$E(t) = k \int_T^t \frac{ds}{r(s)} \tag{I.2}$$

$$\frac{dr}{dt} = \frac{dr}{dE}\frac{dE}{dt} = \frac{k}{r}\frac{dr}{dE}$$

and (I.1) becomes

$$k^2 \left(\frac{dr}{dE} \right)^2 + c^2 = 2(\mu r + h r^2). \tag{I.3}$$

In the *parabolic case,* corresponding to $h = 0$, the choice $k = \sqrt{\mu}$ changes (I.3) to

$$\left(\frac{dr}{dt} \right)^2 + \frac{c^2}{\mu} = 2r. \tag{I.4}$$

With this choice

$$\frac{d^2 r}{dE^2} = 1$$

and (I.4) has a solution

$$r = \frac{1}{2} \left(E^2 + \frac{c^2}{\mu} \right).$$

In (I.2)

$$\frac{dE}{dt} = \frac{k}{r}$$
$$k\,dt = r\,dE$$

$$k(t - T) = \int_0^E \frac{1}{2} \left(s^2 + \frac{c^2}{\mu} \right) ds$$

so that

$$\sqrt{\mu}(t - T) = \frac{1}{6} E^3 + \frac{c^2}{\mu} E. \tag{I.5}$$

This is Barker's equation, with T being the time of perihelion passage.

In the *hyperbolic case,* corresponding to $h > 0$, the choice

$$k = \sqrt{2h} = \sqrt{\frac{\mu}{a}},$$

where a is half the transverse axis, makes (I.3) look like

$$\frac{\mu}{a} \left(\frac{dr}{dt} \right)^2 + c^2 = 2(\mu r + h r^2) = 2\mu r + \frac{\mu}{a} r^2$$

or

$$\left(\frac{dr}{dE} \right)^2 + a^2 e^2 = (a + r)^2 \tag{I.6}$$

since

$$\frac{c^2}{\mu} = a(e^2 - 1).$$

We can see $r = a(e \cosh E - 1)$ satisfies (I.6). Since $k\, dt = r\, dE$, from (I.2)

$$\sqrt{\frac{\mu}{a}}(t - T) = \int_0^E a(e \cosh s - 1)ds$$

and

$$\sqrt{\frac{\mu}{a^3}}(t - T) = e \sinh E - E \tag{I.7}$$

which is the hyperbolic form of KE.

The details of the chain of equations (I.1)–(I.7) are from [Pollard,1976, pp. 16–21].

The eccentric anomaly E, which appears in (I.5) and (I.7), seems rather remote from our original description in Chapter 1. In the parabolic and hyperbolic cases, E may be given a geometric interpretation consistent with the elliptic case but not as an angle. Since "anomaly" in astronomical usage is customarily an angle, calling E an anomaly is a slight abuse of language.

In the parabolic case,

$$\frac{c^2}{\mu} = p,$$

the perihelion distance, and

$$E = \frac{c}{\sqrt{\mu}} \tan\frac{\theta}{2} = \sqrt{p}\tan\frac{\theta}{2}$$

where (r, θ) are the polar coordinates of position at time t.

To compare the hyperbolic case with the elliptic, we describe in Figure I.1 the ellipses

$$\frac{x^2}{a^2} + \frac{y^2}{a^2(1 - e^2)} = 1 \ \text{ and } \ \frac{x^2}{a^2} + \frac{y^2}{a^2} = 1$$

and in Figure I.2 the hyperbolas

$$\frac{x^2}{a^2} - \frac{y^2}{a^2(e^2 - 1)} = 1 \ \text{ and } \ \frac{x^2}{a^2} - \frac{y^2}{a^2} = 1.$$

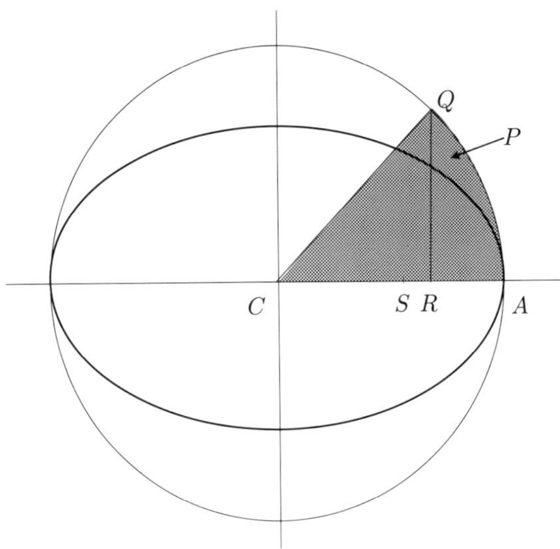

Figure I.1

When P represents position at a given time in the elliptic case, the eccentric anomaly E is given by the angle QCA, and its numerical value equals

$$\frac{1}{a^2} \text{ (Area of sector } QCA\text{).}$$

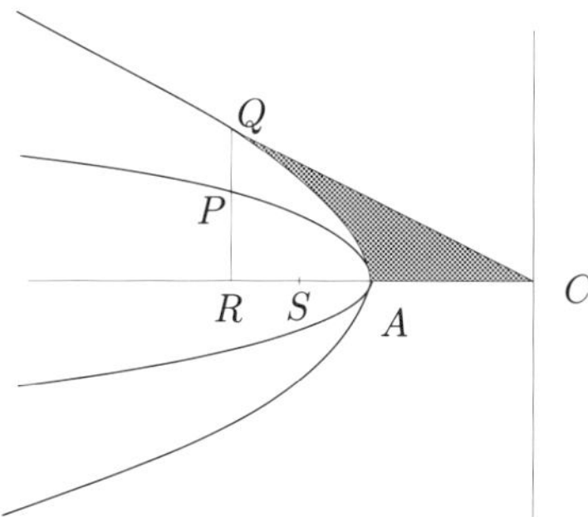

Figure I.2

In the hyperbolic case with all corresponding points marked similarly,

$$E = \frac{1}{a^2} \text{ (Area of shaded region } QCA\text{).}$$

To consider Battin's universal form of KE, suppose at time $t = 0$ an object is at position R_0, has velocity V_0 and eccentric anomaly E_0. Then

$$\text{(a)}\sqrt{\frac{\mu}{a^3}}\,t = (E - E_0) - e(\sin E - \sin E_0),\ 0 < e < 1$$

$$\text{(b)}\sqrt{\frac{\mu}{a^3}}\,t = e(\sinh E - \sinh E_0) - (E - E_0),\ e > 1. \tag{I.8}$$

In the elliptic case

$$e \cos E_0 = 1 - \frac{|R_0|}{a}, \quad e \sin E_0 = \frac{R_0 \cdot V_0}{\sqrt{\mu a}}$$

and we can rewrite (I.8(a)) as

$$\sqrt{\frac{\mu}{a^3}}\,t = (E - E_0) + \frac{R_0 \cdot V_0}{\sqrt{\mu a}}(1 - \cos(E - E_0)) + \left(1 - \frac{|R_0|}{a}\right)\sin(E - E_0). \tag{I.9(a)}$$

In a similar manner, (I.8(b)) may be rewritten as

$$\sqrt{\frac{\mu}{a^3}}\,t = -(E - E_0) + \frac{R_0 \cdot V_0}{\sqrt{\mu a}}(\cosh(E - E_0) - 1) + \left(1 + \frac{|R_0|}{a}\right)\sinh(E - E_0). \tag{I.9(b)}$$

When

$$C(x) = \sum_{n=1}^{\infty} \frac{(-1)^{n-1}x^{n-1}}{(2n)!}, \quad S(x) = \sum_{n=1}^{\infty} \frac{(-1)^{n-1}x^{n-1}}{(2n+1)!}$$

and we substitute

$$\alpha = \frac{1}{a}, \quad E - E_0 = \sqrt{\alpha}\,x$$

in (I.9) we obtain the unified form

$$\sqrt{\mu}\,t = \frac{R_0 \cdot V_0}{\sqrt{\mu}}x^2 C(\alpha x^2) + (1 - |R_0|\alpha)x^3 S(\alpha x^2) + |R_0|x. \tag{UKE}$$

The parabolic case may be viewed as a special limiting case where *perihelion distance*

$$p = \lim_{\substack{a \to +\infty \\ e \to 1}} [a(1 - e)]$$

In (UKE) this limiting case corresponds to $|R_0| = p$, $\alpha = 0$, while $R_0 \cdot V_0 = 0$, so (UKE) becomes a form of Barker's equation

$$\sqrt{\mu}\,t = x^3 S(0) + px = \frac{1}{6}x^3 + px.$$

Appendix J

Boltz's Parameters

Following [Boltz,1986], let e, $0 < e < 1$, be fixed and presume B_1, B_2, B_3, C_2, C_3 are constants for which the equation

$$B_1 E + B_2 E^2 + B_3 E^3 = M + C_2 M^2 + C_3 M^3 \tag{8.11}$$

has a unique solution $E_B(M)$. Also assume $E(M)$ satisfies

$$E(M) - e \sin E(M) = M.$$

In order that $E_B(M) = E(M)$ and $E_B{}'(M) = E'(M)$ for

$$M = 0, \ \frac{\pi}{2} - e, \ \pi,$$

Boltz found that

$$B_1 = 1 - e$$

$$B_2 = \frac{e}{\pi} - \frac{3e}{2D} \left[\pi \left(1 - \frac{\pi}{4} \right) - e(1 - 2e) \right]$$

$$B_3 = \frac{e}{D} \left(1 - \frac{\pi}{4} + \frac{3e^2}{\pi} \right)$$

$$C_2 = \frac{3e}{D} \left[e - \pi \left(1 - \frac{\pi}{4} \right) \right]$$

$$C_3 = \frac{e}{D} \left(1 - \frac{\pi}{4} \right)$$

where

$$D = \left(\frac{\pi}{2} - e\right)\left[(1 + 3e)\frac{\pi}{2} - \left(\frac{\pi}{4} + e\right)\left(\frac{\pi}{2} - e\right)\right]$$

$$- \frac{\left[\left(1 - \frac{\pi}{4}\right)\frac{\pi}{2} + \frac{3e^2}{2}\right]\left[\left(1 - \frac{\pi}{4}\right) - e\right]}{1 - \frac{\pi}{4}}$$

Appendix K

Riemann Boundary Value Problems and the Burniston-Siewert Method

1. Riemann Boundary Value Problems

If Γ is a simple smooth curve in the complex plane $\mathcal{C}$, and if D is any region in $\mathcal{C}$ which Γ divides into two parts, let $D^+(D^-)$ denote the part of D which lies to the left (right) of Γ, relative to the positive direction along Γ.

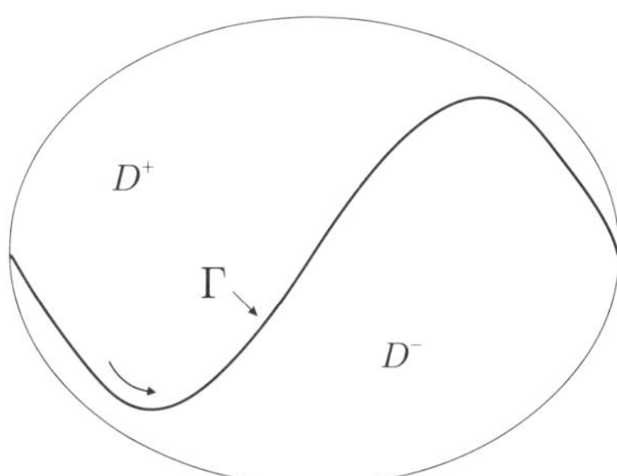

Figure K.1

For any function f defined in $D - \Gamma$ and t on Γ define

$$f^+(t) = \lim_{\substack{z \to t \\ z \in D^+}} f(z)$$

$$f^-(t) = \lim_{\substack{z \to t \\ z \in D^-}} f(z)$$

where these limits exist.

The original boundary value problem of Riemann was this: if Γ is a simple closed curve in $\mathcal{C}$ and D^+ is its interior, and if $a(t)$, $b(t)$, $c(t)$ are nice functions defined on Γ with $a^2(t)+b^2(t) \neq 0$, find a function $f(z) = u(z)+iv(z)$, analytic in D^+, so that

$$a(t)u^+(t) + b(t)v^+(t) = c(t) \tag{R}$$

for each t on Γ.

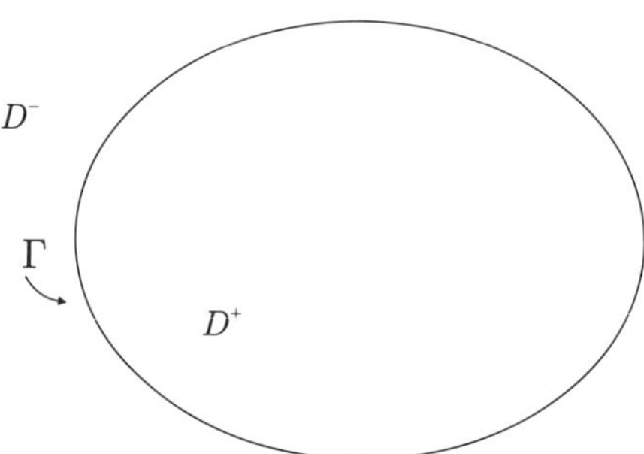

Figure K.2

It was later established by I. Privalov that any solution of (R) also solves the boundary value equation

$$f^+(t) = d(t)f^-(t) + e(t) \tag{P}$$

for t on Γ, where

$$d(t) = -\left[\frac{a(t) - ib(t)}{a(t) + ib(t)}\right]$$

$$e(t) = \frac{2c(t)}{a(t) + ib(t)}.$$

The problem of finding f to satisfy (P) is called a *Privalov boundary value problem*. In the special case where $e(t) = 0$, equation (P) becomes

$$f^+(t) = d(t)f^-(t) \tag{H}$$

and the analogous problem is called a *Hilbert boundary value problem*.

The study of the problem of finding functions f, analytic in $\mathcal{C} - \Gamma$, which satisfy (P) or (H) is well developed. The variety of possible solutions is described by improper line integrals called *Cauchy integrals* or *Hilbert transforms*. An excellent guide for beginning a study of these problems and some of their applications is [Henrici,1986, Chapter 14]. Among these applications is the Burniston-Siewert method for solving certain classes of transcendental equations.

If Γ is an arc in $\mathcal{C}$, and if we seek f analytic in $\mathcal{C} - \Gamma$ to satisfy (H), it is known that, subject to certain technical hypotheses on f and $d(t)$,

a) f will have only finitely many zeros;

b) a polynomial may be constructed whose zeros are the same as those of f;

c) the construction will depend on a finite number of the Laurent coefficients of

$$f\left(\frac{1}{z}\right)$$

and the quantities

$$\left\{ \begin{array}{l} \dfrac{1}{2\pi i} \oint_\Gamma t^k \log d(t)\,dt \\[2mm] k = 0, 1, 2, 3, \ldots \end{array} \right\} \tag{K.1}$$

2. The Burniston-Siewert Method and KE

The Burniston-Siewert method achieves its results by recasting the solution of a transcendental equation as the solution of a Hilbert boundary value problem for a suitable arc Γ and boundary function $d(t)$. With careful analysis based on the zero-counting machinery of Cauchy integral theory, but analysis which is *ad hoc* in the sense it may vary with the original equation being treated, the method leads to polynomials whose roots are branches of the solution function and are expressed in terms of integrals of the form (K.1) and Cauchy integral form. The article [Burniston,1973] illustrates the method and discusses conditions sufficient for its use.

As is the case for other techniques discovered for the analytical treatment of integral equations involved in boundary value problems—the Wiener-Hopf technique, for example—the Burniston-Siewert method requires theoretical knowledge, technical skill, and insights that not all who would use the method are able to control. We'll be content here to show how KE may be transformed into a Hilbert problem and leave further steps to the references cited above.

With

$$a = \frac{1}{e}, \quad b = \frac{M}{e}, \quad z = \frac{e}{E - M},$$

KE may be written in the form

$$F(z) = 1 + bz - a \sin^{-1}\left(\frac{1}{z}\right) = 0. \tag{K.2}$$

For each integer k there is a branch of

$$\sin^{-1}\left(\frac{1}{z}\right)$$

analytic in $\mathcal{C} - \{t : -1 \le t \le 1\}$ given by

$$k\pi + (-1)^k \left[\frac{\pi}{2} - \mathrm{Log}\left(\sqrt{\frac{1}{z^2} - 1} + \frac{1}{z}\right)\right] \tag{K.3}$$

where it is understood that the principal square root and logarithm function are indicated. We denote by $F_k(z)$ the version of (K.2) which uses the branch of

$$\sin^{-1}\left(\frac{1}{z}\right)$$

from (K.3) corresponding to the choice of k.

Solutions of E in KE may be determined from the solutions $\{z_{kj}\}$ of (K.2) for each k.

Let $\Gamma = \{t : -1 \le t \le 1\}$ and $F_k(t)$ denote the boundary values of F_k relative to Γ. Define

$$G_k(z) = F_k(z)F_k(-z) \tag{K.4}$$

and

$$H_k(t) = \frac{G_k^+(t)}{G_k^-(t)}, \quad 0 < t < 1. \tag{K.5}$$

The Hilbert boundary value problem Burniston and Siewert used for KE was this:

Find $f_k(z)$ analytic in $\mathcal{C} - \Gamma$ so that

$$\tag{HBVP}$$

$$f_k^+(t) = H_k(t)f_k^-(t).$$

With k chosen, let $2\pi n_k$ be the net increase in arg $H_k(t)$ as t increases from 0 to 1. The canonical solution of (HBVP) is

$$f_k(z) = (1 - z)^{-n_k} \exp\left[\frac{1}{\pi} \int_0^1 \frac{\arg G_k^+(t)}{t - z} dt\right] \tag{K.6}$$

which has no zeros in $\mathcal{C} - \Gamma$. All other solutions of (HBVP) must be polynomial multiples of this canonical solution. Since, from the definition of F_k and (K.4) it follows that

$$G_k(z) = G_k(-z) \quad \text{and} \quad \overline{G_k(\bar{z})} = G_k(z)$$

it is also true that

$$\frac{G_k(z)}{f_k(-z)}$$

solves (HBVP). So there is a polynomial $P_k(z)$ of degree $2(n_k + 1)$ such that

$$\frac{G_k(z)}{f_k(-z)} = f_k(z)P_k(z). \tag{K.7}$$

In the elliptic case, $0 < e < 1$, only real solutions of KE are sought; the zeros of $G_k(z)$, which are directly related to those of $F_k(z)$, thus come in pairs z_{kj}, $-z_{kj}$ because of (K.4). This means the polynomial $P_k(z)$ must consist of factors $z^2 - z_{kj}^2$ and the number of allowable (k, n_k)-pairs is small. In fact, only the pairs $(1, 2)$, $(0, 0)$, $(3, 2)$ are allowable, and these pairs correspond to the (e, M)-regions R_1, R_2, R_3 described in Figure 8.1.

The remainder of the method applied to KE consists of describing the zeros of $P_k(z)$ in terms of $F_k(z)$ and (K.6) and expressing

$$E = M + \frac{e}{z}.$$

Appendix L

Newton's Method for Formal Power Series

Following the language and notation of Chapter 8, suppose f and g lie in $\wp$. The equation

$$(f \circ w)(x) = g(x) \tag{L.1}$$

has a unique solution in $\wp$ given by

$$w(x) = (f^{[-1]} \circ g)(x). \tag{L.2}$$

If $(w)_k(x)$ is an approximation of $w(x)$, then $(w)_k$ lies in $\wp$ and for some fps v

$$(w)_k = w + x^k v.$$

If we use Newton's method in the customary sense to solve (L.1) with initial approximation $(w)_k$, then

$$u = (w)_k - \frac{(f \circ (w)_k) - g}{f' \circ (w)_k}$$

is taken as the next approximation for any x.

In fact, for some fps z

$$u = w + x^{2k} z$$

so that u has twice as many correct coefficients of w as $(w)_k$. Linear approximation will explain why:

$$
\begin{aligned}
f \circ (w)_k &= f \circ (w + x^k v) \\
&\approx f \circ w + (f' \circ w)(x^k v) + O(x^{2k}) \\
&= g + (f' \circ w)(x^k v) + O(x^2 k).
\end{aligned}
$$

155

Also

$$f' \circ (w)_k = f' \circ (w + x^k v)$$
$$\approx f' \circ w + (f'' \circ w)(x^k v)$$
$$= f' \circ w + O(x^k)$$

so that

$$u = (\dot{w})_k - \frac{f \circ (w)_k - g}{f' \circ (w)_k}$$
$$\approx (w)_k - \frac{[(f' \circ w)(x^k v) + O(x^{2k})]}{[f' \circ w + O(x^k)]}$$
$$= w + x^k v - x^k v \left(\frac{f' \circ w + O(x^k)}{f' \circ w + O(x^k)} \right)$$
$$= w + x^k v - x^k v[1 + O(x^k)]$$
$$= w + O(x^{2k}).$$

The algorithm suggested to solve (L.1) by these steps is this:

Set
$$u_0 = 0$$
$$u_{m+1} = u_m - \left(\frac{f \circ u_m - g}{f' \circ u_m} \right)_{2m+1}$$
$$m = 0, 1, 2, \ldots$$
Then
$$u_{m+1} = (w)_{2m+1}$$

Fuller details are to be found in [Henrici,1986, pp. 71–78], and some examples appear in [Kung,1978]. As our illustration we'll follow the algorithm for several steps on KE.

With e fixed, let $E(M)$ be the solution of KE; let $f(x) = x - e\sin x$ and $g(x) = x$ so that f and g lie in $\wp$. Then

$$(f \circ E)(M) = g(M).$$

If $u_0 = 0$,

$$u_1 = u_0 - \left(\frac{f \circ u_0 - g}{f' \circ u_0} \right)_2$$
$$= 0 - \left(\frac{0 - M}{1 - e} \right)_2$$
$$= \frac{M}{1 - e}$$

$$u_2 = u_1 - \left(\frac{f \circ u_1 - g}{f' \circ u_1} \right)_4$$

$$= \frac{M}{1-e} - \left(\frac{f\left(\frac{M}{1-e}\right) - M}{1 - e\cos\left(\frac{M}{1-e}\right)} \right)_4$$

$$= \frac{M}{1-e} - \left[\frac{\frac{M}{1-e} - e\sin\left(\frac{M}{1-e}\right) - M}{1 - e\cos\left(\frac{M}{1-e}\right)} \right]_4$$

$$= \frac{M}{1-e} - \left[\frac{\frac{M}{1-e} - e\left\{ \frac{M}{1-e} - \frac{1}{3!}\left(\frac{M}{1-e}\right)^3 + \ldots \right\} - M}{1 - e\left\{ 1 - \frac{1}{2!}\left(\frac{M}{1-e}\right)^2 + \frac{1}{4!}\left(\frac{M}{1-e}\right)^4 - \ldots \right\}} \right]_4$$

$$= \frac{M}{1-e} - \left[\frac{\frac{e}{3!}\left(\frac{M}{1-e}\right)^3 - \frac{e}{5!}\left(\frac{M}{1-e}\right)^5}{(1-e) + \frac{e}{2!}\left(\frac{M}{1-e}\right)^2 - \frac{e}{4!}\left(\frac{M}{1-e}\right)^4} \right]_4$$

$$= \frac{M}{1-e} - \frac{eM^3}{3!(1-e)^4}$$

$$u_3 = u_2 - \left(\frac{f \circ u_2 - g}{f' \circ u_2} \right)_8.$$

We'll not write the details; expanding these expressions in powers of M, dividing, and keeping powers of M only through M^8,

$$u_3 = \frac{M}{1-e} - \frac{eM^3}{3!(1-e)^4} + \frac{e(9e+1)}{5!(1-e)^7}M^5$$

$$- \frac{e(225e^2 + 54e + 1)}{7!(1-e)^{10}}M^7.$$

The expressions

$$\left(E(M)\right)_2 = \frac{M}{1-e}, \quad \left(E(M)\right)_4 = u_2, \quad \left(E(M)\right)_8 = u_3$$

are partial sums of the Lie-series solution of Stumpff.

Appendix M

References Sorted by Categories

See pages 169–170 for description of categories.

Category	Subcategory	Author	Date
A		Flauti	1857
		Scorza	1857
	1	Boulliau	1645
		Brenke	1936
		Cassini	1669
		Gaythorpe	1920
			1925
			1957
		Grunert	1856
		Howe	1879
			1880a,b
			1884
			1919
			1922
		Jorgensen	1974
		Marth	1890a,c
		Mercator	1664
		Oppolzer	1885
		Plummer	1919
		Radau	1882
		Schering	1884
		Wallace	1835
			1836
		Willis	1929
		Zenger	1882a,b,e
	2	Cassini	1721
		De la Caille	1750a,b
		Ellicott	1799
		Encke	1850

Category	Subcategory	Author	Date
		Hermann	1726
		Herz	1881
		Mercator	1670
		Robertson	1816
		Stewart	1756
	3	Keill	1713
B	1	Adams	1849
			1882
		Dolainski	1913
		Dubois	1863
		Moulton	1895
			1930
		Proctor	1873
		Radau	1894
		See	1896
		Waterston	1850
		Zelbr	1895
	2	Gonggrijp	1901
		Wallis	1695
	3	Bazenow	1929
C		Barton	1979
		De Gasparis	1882
		Delambre	1820
		LeVerrier	1855
		Lindqvist	1778
		Rohde	1806
		Weiler	1899
		Weiss	1881
	1	Kapteyn	1883
	2	Adams	1883
		Bryant	1887
		Camerer	1805
		Charlier	1927
		Creedy	1857
		Degen	1821
			1823
		Fleckenstein	1941
		Glaisher	1877b
		Klugel	1789
		Lagrange	1770
			1771
		Laplace	1966
		Littrow	1818

Category	Subcategory	Author	Date
		Morrison	1883
		Moulton	1907
		Picart	1906
		Puiseux	1884
		Rouché	1862
		Schubert	1818
			1820a,b
		Tchebychef	1857a,b
		Weiss	1884
	3	Bessel	1818
			1819
		Bourget	1861
		Carlini	1817
			1818
		Cauchy	1844
			1854b,c
		Cayley	1858
		Greathead	1837
		Hall	1905
		Hansen	1852a,b,c
			1853a,b,c
			1855
		Hennert	1804
		Innes	1924
		Jacobi	1836
			1848
			1850
		Kapteyn	1893
		Laska	1884
		LeFort	1846
		Oppolzer	1878
		Poisson	1811
			1825
		Posselt	1818
		Scheibner	1856a,b
			1880a,b
		Sconzo	1933b
			1934
	4	Bourget	1854
		Jekhowsky	1918
	5	Cauchy	1841
	6	Kaminskii	1982
		Levi-Civita	1904a,b
	7	DeGasparis	1879a
		Moulton	1903

Category	Subcategory	Author	Date
		Paul	1963
		Rittenhouse	1799
		Stumpff	1968b
	8	Block	1904
		Callendreau	1881
		Cauchy	1829a,b
			1831
			1840a,b
			1843
			1857
		Laplace	1823
		Puiseux	1849
			1860
		Serret	1856
			1857
			1859
D		Sinnott	1985
	1	DeGasparis	1881c
		Euler	1740
			1748
			1750
		Fuss	1788
		Grunert	1851
		Karlinski	1862a
		Marth	1890b
		Pacassi	1794
		Peters	1891a
		Trembley	1782
		Wolfers	1862a,b
	2	Colwell	1991
		Newton	1934
		Ng	1979
		Nicollic	1746
		Schubert	1854
		Sconzo	1933a
		Smith, G.	1979
		Smith, O.K.	1968
		Taylor	1988
	3	Ivory	1805
	4	Battin	1979
		Benima	1969
		Fill	1976
		Gauss	1809

Category	Subcategory	Author	Date
		Karlinski	1862b
		Wolfers	1865
		Zenger	1882d
E		Bergam	1980
			1982
		Estes	1970
		Gedeon	1964
		Gooding	1987b
		Pitkin	1965a,b
		Prussing	1979
		Sarnecki	1988
		Shepperd	1985
			1987
		Stumpff, K.	1968a
		Stumpff, P.	1987
F	1	Bina	1990
		Broucke	1980
		Cherry	1955
		Dass	1990
		Gourier	1878
		Hagihara	1936
		Ioakimidis	1985
		Langebartel	1948
		Neutsch	1986
		Siewert	1972
	2	Bobone	1930
		Boltz	1986
			1987
		Brown	1931
		Conway	1986
		Danby	1987
		DeGasparis	1856
		De la Hire	1713
		Grandjean de Fouchy	1738
			1740
		Hartmann	1918
		Klugel	1806
		Machin	1738
		Mikkola	1987
		Nijenhuis	1990
		Peters	1984
		Plummer	1903
			1906
		Rust	1908
		Sheela	1982
		Wilkens	1918

Category	Subcategory	Author	Date
		Witt	1906
	3	Glaisher	1877a
		Gylden	1875a,b
G		Andoyer	1918
		Cronstrand	1813
		Davidson	1933
			1934
		Herrick	1945
		Moller	1933
			1940
		Oppolzer	1879
			1880
		Searle	1889
			1892
		Smiley	1934
	1	Barker	1757
		Bessel	1845
		Halley	1705
		Herget	1936
		Innes	1932
		Nicolai	1844
		Olbers	1797
	2	Simpson	1740
	3	Bessel	1805
		Brunnow	1858
H	1	Herglotz	1910
		Houzeau	1882
		Knobel	1876
		Radau	1900
		Struve	1860
		Wood	1950a,b
	2	Brinkley	1803
		Burkhardt	1983
		Christom	1744
		Colwell	1992
		Danby	1983
		Detmoldt	1798
		Gooding	1985
		Karlinski	1862c
		Littrow	1824
		Melander	1767
		Mossotti	1817a
		Nordmark	1799
		Odell	1986

Category	Subcategory	Author	Date
		Radau	1885
		Seydler	1887
		Sorokin	1982
		Taff	1989
		Villarceau	1857
		Voll	1826
	3	Belyi	1975
		Bottazzini	1986
		Boyer	1989
		Cajori	1911
		Caspar	1932
		Grabiner	1981
		Iushkevich	1966
			1976
		Jackson	1988
		Kennedy	1956
		Kennedy & Transue	1956
		Kline	1972
		Moore	1987
		Pagels	1985
		Russell	1964
		Stephenson	1987
		Valson	1868
		Varignon	1705
		Whiteside	1965
			1968
			1974
		Wilson	1968
			1970
			1989
		Wolfers	1846
			1863
	4	Agnew	1942
		Barbeau	1979
		Beckmann	1971
		Bessel	1826
		Birkhoff	1973
		Brill	1894
		Brinkley	1800
		Burkhardt	1908
		Burniston	1973
		Cauchy	1854d,e
		Delves	1967
		Grattan-Guiness	1972
		Halley	1694
		Henrici	1974
			1986
		Hurwitz	1906

Category	Subcategory	Author	Date
		Kline	1983
		Kung	1978
		Osgood	1905
		Speiser	1947
		Watson	1922
			1948
		Whittaker	1927
		Wintner	1941
	5	Bailey	1990
		Ball	1915
		Battin	1964
		Bessel	1836
		Broucke	1973
		Camerer	1794
		D'Amario	1969
		Danby	1988
		Duffet-Smith	1981
		Hauff	1797
		Herrick	1971
		Lehmann	1855
			1856
		Moulton	1914
		Nordmark	1789
		Oppolzer	1882
		Perrey	1853
		Pollard	1976
		Roy	1988
		Taff	1985
		Thiele	1883
		Weiler	1884
		Wilkens	1931
		Wolfers	1849
	6	Bertrand	1870
		Delambre	1812
		Kreutz	1880
		Littrow	1815
		Peters	1891b
		Radau	1888
I		Aldis	1902
		Andoyer	1918
		Astrand	1887
			1890
		Bessel	1845
		Callendreau	1885
		Cayley	1861
		DeGasparis	1881a

Category	Subcategory	Author	Date
		Doberck	1878
		Draper	1936
		Herget	1936
		Herz	1881
		Innes	1933
		Kletke	1852
		Lehmann	1857
		Marth	1890b
		Moller	1933
			1940
		Nicolai	1844
		Oppolzer	1879
			1880
			1885
		Searle	1889
			1892
		Seydler	1888
		Udick	1912
		Viaro	1920
		Wolfers	1862a,b
		Zenger	1882c,d
J	1	Carlini	1856
		Ueta	1927
	2	Callendreau	1879
		Dazew	1934
		Plummer	1896
			1907
		Rambaut	1890
			1906
			1913
		Wilczynski	1912
			1913
	3	DeGasparis	1857a,b,c,d,f
			1858
			1862a,b,c
			1863
			1877a,b,c
			1881b
		Levin	1922
		Mache	1910
	4	Moulton	1895
			1930
		Pritchard	1877
		See	1895

Category	Subcategory	Author	Date
	5	Brodetsky	1938
		D'Ocagne	1894
		Duffet-Smith	1981
		Fry	1916
		Laska	1914
		Le Grand Roy	1903
		Radau	1883
			1884
		Rand	1942
		Wood	1946
K		Bergam	1980
			1982
		Boltz	1986
			1987
		Broucke	1980
		Cassini	1721
		Conway	1986
		Danby	1987
		Ellicott	1799
		Encke	1850
		Gooding	1987a
			1988
		Hermann	1726
		Howe	1922
		Jorgensen	1974
		Kapteyn	1883
		Karlinski	1862a
		Klumpp	1990
			1991
		Machin	1738
		Ng	1979
		Nijenhuis	1990
		Prussing	1977
			1979
		Robertson	1816
		Serafin	1986
		Smith	1979
		Stewart	1756
		Taff	1989
		Taylor	1988
M	1	Bina	1990
	2	Ioakimidis	1985
		Neutsch	1986
		Siewert	1972
	3	Sheela	1982

Bibliography

The bibliography items are listed alphabetically by author, and entries for each author are listed chronologically.

In addition to entries which directly bear on Kepler's Equation, the bibliography contains citations for works we found useful or relevant for background.

Each entry directly relevant to Kepler's Equation is marked at its end with a classification code and, if relevant, a number indicating the page on which the entry is first cited. The classification is based on eight primary categories and five secondary categories designated by letters A–M, and these categories may have subcategories denoted by numbers. For an entry involving Kepler's Equation, a typical code may look like

$$A3J1 \ [124]$$

indicating that the entry lies in primary category A, subcategory 3; lies in secondary category J, subcategory 1; is first cited on page 124.

While this classification may seem hopelessly mathematical and cumbersome, the collection of categories has been chosen economically, and the labeling of entries gives considerable information.

The table of categories and subcategories appears on the next page. In Appendix M we present a list of all classified entries sorted according to categories and subcategories and identified by author and date.

PRIMARY CATEGORIES

A. GEOMETRIC
  1. Empty focus
  2. Cassini
  3. Newton

B. GRAPHICAL
  1. Curve of sines
  2. Cycloid
  3. Nomograph

C. INFINITE SERIES
  1. Keill
  2. Lagrange
  3. Bessel-Fourier
  4. Generalized Bessel
  5. Cauchy
  6. Levi-Civita
  7. Stumpff
  8. Convergence and 0.662743...

D. ITERATION
  1. Kepler
  2. Newton
  3. Ivory
  4. Gauss

E. UNIVERSAL FORM

F. DIVERSE METHODS
  1. Analytic function theory
  2. Empirical model
  3. Elliptic functions

G. PARABOLIC APPROXIMATION
  1. Barker's equation
  2. Simpson's method
  3. Bessel's improvement

H. BACKGROUND
  1. Bibliography
  2. Survey article
  3. History
  4. Mathematics
  5. Celestial Mechanics
  6. Review article

SECONDARY CATEGORIES

I. AUXILIARY DEVICE WITH
   INTERPOLATION

K. REFINED STARTING VALUE
   WITH ITERATION

J. DEVICE
  1. Cassini
  2. Cycloid
  3. Linear scale/slide rule
  4. Curve of sines
  5. Nomograph

M. NUMERICAL ALGORITHM
  1. Polynomial solver
  2. Evaluation of integrals
  3. Linear programming

Adams, J.C., 1849. "On the application of graphical methods to the solution of certain astronomical problems, and in particular to the determination of the perturbations of Planets and Comets," *Reports of the British Association for the Advancement of Science.* B1[21]

Adams, J.C., 1877. "On a simple proof of Lambert's Theorem," *Messenger of Mathematics,* **7,** 97–100.

Adams, J.C., 1882. "On Newton's solution of Kepler's Problem," *Monthly Notices of the Royal Astronomical Society,* **43,** 43–49. B1[48]

Adams, J.C., 1883. "Note on Dr. Morrison's paper on the computation of the eccentric anomaly," *Monthly Notices of the Royal Astronomical Society,* **43,** 365–368. C2

Agnew, R., Higgins, T. and Rosser, B., 1942. "Note on Whittaker's method for the roots of a power series," *The American Mathematical Monthly,* **49,** 462–465. H4[94]

Aldis, W.S., 1902. "Explanation of the use of the tables of $(1/2)(\Theta + \cos\Theta)$," *Monthly Notices of the Royal Astronomical Society,* **62,** 633–643. I[84]

Andoyer, H., 1918. "Formules et tables nouvelles relatives à l'étude du mouvement des comètes et à différents problèmes de la théorie des orbites," *Bulletin Astronomique,* **35,** 5–51. GI[94]

Astrand, J.J., 1887. "Om en auxiliaertabel til loesning af Kepler's Problem," Mindre Afhandlingar. I

Astrand, J.J., 1890. *Hulfstafeln zur leichten und genauen Auflosung des Kepler'schen Problemes,* Verlag von W. Engelmann, Leipzig. I[84]

Bailey, M., Clube, S. and Napier, W., 1990. *The Origin of Comets,* Pergamon Press, Oxford. H5[79]

Ball, R., 1915. *A Treatise on Spherical Astronomy,* Cambridge University Press, Cambridge, 156–158. H5

Barbeau, E., 1979. "Euler subdues a very obstreperous series," *The American Mathematical Monthly,* **86,** 356–372. H4[24]

Barker, T., 1757. "An account of discoveries concerning comets, with the way to find their orbits, and some improvements in constructing and calculating their places," London. G1[58]

Barton, D., 1979. "Expansions in power series by computer," *Celestial Mechanics,* **20,** 3–18. C

Battin, R.H., 1964. *Astronomical Guidance,* McGraw-Hill, New York, Chapter 2. H5[96]

Battin, R.H. and Fill, T.J., 1979. "Extension of Gauss' method for the solution of Kepler's equation," *Journal of Guidance and Control,* **2,** 190–195. D4

Bazenow, G., 1929. "Graphische Losung einiger Gleichungen der theoretische Astronomie," *Astronomische Nachrichten,* **234,** 425–426. B3

Beckmann, P., 1971. *A History of Pi,* St. Martin's Press, New York. H4[23]

Belyi, Yu.A., 1975. "Johannes Kepler and the Development of Mathematics," in "Kepler: Four Hundred Years," Beer and Beer, eds., *Vistas in Astronomy,* **18,** 643–660. H3

Benima, B., Cherniak, J. and Marsden, B., 1969. "The Gauss method for solving Kepler's equation in nearly parabolic orbits," *Publications of the Astronomical Society of the Pacific,* 121–129. D4[66]

Bergam, M.J., 1980. "New Bounds and Efficient Starting Values for the Universal Kepler's Equation," M.S. Thesis, University of Illinois. EK[101]

Bergam, M.J. and Prussing, J.E., 1982. "Comparison of starting values for iterative solutions to a universal Kepler's equation," *Journal of the Astronautical Sciences,* **30,** 75–84. EK

Bertrand, J., 1870. Review of C.A. Valson, "La vie et les travaux du Baron Cauchy," *Bulletin des Sciences Mathématiques et Astronomiques,* **1,** 105–117. H6

Bessel, F.W., 1805. "Ueber die Berechnung der wahren Anomalie in einer von der Parabel nicht sehr verscheidenen Bahn," *Monatliche Correspondenz zur Beforderung der Erd-und Himmels-Kunde herausgegeben von Freiherrn von Zach,* **12,** 197–207. G3[29]

Bessel, F.W., 1818a. "Letter to W. Olbers of 23 April 1818," *Briefwechsel zwischen W. Olbers und F.W. Bessel,* A. Erman, ed., Avenarius and Mendelssohn, Leipzig, **2,** 84–90. C3[30]

Bessel, F.W., 1818b. "Ueber die Entwickelung der Mittels-puncktsgleichung und des Radiusvectors in Reihen," *Zeitschrifte für Astronomie,* **5,** 367–375. C3[29]

Bessel, F.W., 1819. "Analytische Auflosung der Kepler'schen Aufgabe," *Abh. Akad. Wiss. Berlin, math. Kl.,* 49–55. C3[29]

Bessel, F.W., 1826. "Untersuchung des Theils der planetarischen Storungen welcher aus der Bewegung der Sonne ensteht," *Abh. Akad. Wiss. Berlin, math. Kl.,* 1–52. H4[29]

Bessel, F.W., 1836. "Bemerkungen uber mogliche Unzulangkeit durche Anzeilungen allein berucksichtigenden Theorie der Kometen," *Astronomische Nachrichten,* **13,** 345–350. H5

Bessel, F.W., 1845. "Ueber die Berechnung der wahren parabolische Anomalie aus der Zeit, fur Falle, in weichen sie sich 180° nahert," *Astronomische Nachrichten,* **22,** 253–256. G1I

Bina, S., 1990. "The Delves-Lyness solution," M.S. paper, Department of Mathematics, Iowa State University. F1M1[108]

Birkhoff, G., 1973. *A Source Book in Classical Analysis,* Harvard University Press, Cambridge, Chapters 1–3. H4[67]

Block, H.G., 1904. "Die singuläre Curve der hyperbolischen Bewegung," *Arkiv for Matematik, Astronomi och Fysik,* **1,** 467–479. C8

Bobone, J., 1930. "Resolucion de la ecuacion de Kepler," *Revista Astronomica,* **2,** 398–402. F2[94]

Boltz, F.W., 1986. "Efficient inverse solution of Kepler's equation," *Journal of the Astronautical Sciences,* **34,** 431–443. F2K[103]

Boltz, F.W., 1987. "Inverse solution of Kepler's equation for hyperbolic orbits," *Journal of the Astronautical Sciences,* **35,** 347–358. F2K[103]

Bottazzini, U., 1986. *The Higher Calculus: A History of Real and Complex Analysis from Euler to Weierstrass,* Springer-Verlag, Berlin. H3[75]

Boulliau, I., 1645. *Astronomia philolaica.* A1[9]

Bourget, J., 1854. "Mémoire sur le développement des coordonnées d'une planète en fonction du temps," *Comptes Rendus (Paris)*, **38**, 807–810. C4

Bourget, J., 1861. "Mémoires sur les nombres de Cauchy et leur application à divers problèmes de mécanique céleste," *Journal de Mathématiques pures et appliquées* (2), **6**, 33–54. C3[34]

Boyer, C.B. and Merzbach, U.C., 1989. *A History of Mathematics,* 2nd Ed., Wiley, New York. H3[20]

Brag, J., 1818. *De anomalia vera indirecte invenienda,* Lund.

Brenke, W.C., 1936. "An angle connected with the mean place in the ellipse," *Popular Astronomy,* **44**, 76–78. A1[7]

Brill, A. and Noether, M., 1894. "Entwickelung der Theorie der algebraischen Functionen in alterer und neuerer Zeit," *Jahresbericht der Deutsche Mathematiker-Vereinigung,* **3,** Chap. II. H4[68]

Brinkley, J., 1800. "A method of expressing, when possible, the value of one variable quantity in integral powers of another and constant quantities, having given Equations expressing the relation of these variable quantities; in which is contained the general doctrine of Reversion of Series, of approximating to the Roots of Equations, and of the solution of fluxional Equations by Series," *Transactions of the Royal Irish Academy,* **7,** 321–356. H4

Brinkley, J., 1803. "Examination of various solutions of Kepler's problem, and a short practical solution of that problem pointed out," *Transactions of the Royal Irish Academy,* **9,** 83–131. H2[9]

Brodetsky, S., 1938. *A First Course in Nomography,* Bell, London. J5[157]

Broucke,R., 1980. "On Kepler's equation and strange attractors," *Journal of the Astronautical Sciences,* **28,** 255–265. F1K[100]

Broucke, R. and Cefola, P., 1973. "A note on the relations between true and eccentric anomalies in the two-body problem, *Celestial Mechanics,* **7,** 388–389. H5

Brown, E.W., 1931. "On a method of solving Kepler's equation," *Monthly Notices of the Royal Astronomical Society,* **92,** 104. F2[86]

Brunnow, F., 1858. "On the calculation of the true anomaly in ellipses and hyperbolas of great excentricity," *Astronomical Notices,* 12–14. G3[65]

Bryant, R., 1887. "On Kepler's problem," *Monthly Notices of the Royal Astronomical Society,* **47,** 8–11. C2[25]

Burkhardt, H., 1908. "Die Entwickelung analytischer Functionen in harmonische trigonometrische Reihen, Chapter III in Entwickelungen nach oscillirenden Functionen und Integralen der Differential-Gleichungen der Mathematischen Physik," *Jahresbericht der Deutsche Mathematiker-Vereinigung,* **10,** 94–111. H4[30]

Burkhardt, T. and Danby, J., 1983. "The solution of Kepler's equation, II," *Celestial Mechanics*, **31**, 317–328. H2[102]

Burniston, E. and Siewert, C., 1973. "The use of Riemann problems in solving a class of transcendental equations," *Proceedings of the Cambridge Philosophical Society*, **73**, 111–118. H4[109]

Cajori, F., 1911. "Historical note on the Newton-Raphson method of approximation," *The American Mathematical Monthly*, **18**, 29–32. H3[49]

Callendreau, O., 1879. "Sur le choix de la fonction du temps sous les signes sinus et cosinus dans les expressions des Perturbations," *Vierteljahrbuch der Astronomische Gesellschaft*, **14**, 402–407. J2[88]

Callendreau, O., 1881. "Contribution à la théorie du mouvement elliptique," *Journal de l'Ecole Polytechnique*, **30**, 197–204. C8

Callendreau, O. and Fabry, L., 1885. "Tables numériques destinées à faciliter le calcul des éphémérides des petites planètes," *Bulletin Astronomique*, **2**, 453–464. I

Camerer, J.W., 1794. "Zwei Formeln, die wahre Anomalie eines Planeten aus der mittleren zu berechnen," *Astronomisches Jahrbuch*, 118–119. H5

Camerer, J.W., 1805. "Mit welchem Grade der Genauigkeit lasst sich die grosste Mittelpunkts-Gleichung eines Planeten aus seiner Excentricitat, und umgekehrt diese aus jener vermittelst Reihen bestimmen?," *Astronomisches Jahrbuch*, 147–158. C2

Carlini, F., 1817. "Richerche sulla convergenze della serie che serve alla soluzione del Problemo di Keplero," *Giornale de Fisica, Chimica, e Storio Naturale*, **10**, 458–460. C3[32]

Carlini, F., 1818. "Richerche sulla convergenze della serie che serve alla soluzione del Problemo di Keplero," *Effemeridi astronomiche di Milano*, Appendice, 3–48. C3[32]

Carlini, F., 1856. "Descrizione d'une macchinetta che serve a risolvere il problema di Keplero (ossia a trovare l'anomalia data l'anomalia media qualunque sia l'eccentricita," *Memorie dell'I.R. Instituto Lombardo di Scienze, Lettere ed Arti*, **5**, 93–100. J1[15]

Caspar, M., 1932. "Kepler und die Infinitesemalrechnung," *Unterrichtsblatter fur Math. u. Naturw.*, **38**, 227–229. H3

Cassini, J., 1721. "Méthode de déterminer la premiere Equation des Planètes suivant l'hypothese de Kepler," *Mémoires de l'Académie Royale des Sciences (Paris)*, 147–156. A2K[12]

Cassini, J.-D., 1669. "Nouvelle manière géométrique et directe de trouver les apogées, les excentricités, et les anomalies du mouvement des planètes," *Mémoires de l'Académie Royale des Sciences (Paris)*, **10**, 488–491. A1[11]

Catalan, E., 1884. "Note sur le théoreme de Lambert," *Nouv. annal. d. math. (3)*, **3**, 506–513.

Cauchy, A., 1829a. "Mémoire sur divers points d'analyse," *Mémoires de l'Académie Royale des Sciences (Paris)*, **8**, 97–129. C8[69]

Cauchy, A., 1829b. "Mémoire sur le développement de $f(y)$ suivant les puissances ascendantes de h, y étant une racine de l'équation $z - x - hw(z) = 0$," *Mémoires de l'Académie Royale des Sciences (Paris)*, **8**, 130–138. C8[72]

Cauchy, A., 1831. "Mémoire sur le Mécanique Céleste et sur un nouveau calcul appelé Calcul des Limites," *Mémoires de l'Académie Royale des Sciences (Turin)*. C8[72]

Cauchy, A., 1840a. *Exercises d'Analyse et de Physique Mathématique*, Bachelier, Paris. C8[69]

Cauchy, A., 1840b. "Méthodes générales pour la détermination des mouvements des planètes et de leurs satellites," *Comptes Rendus (Paris)*, **11**, 179–184. C8

Cauchy, A., 1841. "Méthodes propres à simplifier le calcul des inégalités périodiques et séculaires des mouvements des planètes," *Comptes Rendus (Paris)*, **12**, 84–101. C5[72]

Cauchy, A., 1843. "Mémoire sur l'application du calcul des limites à l'astronomie," *Comptes Rendus (Paris)*, **17**, 1157–1159. C8

Cauchy, A., 1844. "Note sur l'application de la méthode logarithmique à la détermination des inégalités périodiques des mouvements planétaires," *Comptes Rendus (Paris)*, **19**, 159–162. C3

Cauchy, A., 1854a. "Formules générales pour la transformation des fonctions implicites en fonctions explicites," *Comptes Rendus (Paris)*, **38**, 945–952. H4[73]

Cauchy, A., 1854b. "Applications des formules établies dans le précéndent Mémoire à la solution des problèmes astronomiques," *Comptes Rendus (Paris)*, **38**, 952. C3[73]

Cauchy, A., 1854c. "Sur la transformation des variables qui déterminent les mouvements d'une planète ou même d'une comète en fonction explicite du temps, et sur le développement de ces fonctions en séries convergentes," *Comptes Rendus (Paris)*, **38**, 990–993. C3

Cauchy, A, 1854d. "Sur les services que la spirale logarithmique peut rendre à l'astronomie," *Comptes Rendus (Paris)*, **38**, 1033–1035. H4

Cauchy, A., 1854e. "Sur la résolution des équations et sur le développement de leurs racines en séries convergentes," *Comptes Rendus (Paris)*, **38**, 1104–1107. H4

Cauchy, A., 1857. "Sur les compteurs logarithmiques appliqués au dénombrement et la séparation des racines des équations transcendantes," *Comptes Rendus (Paris)*, **44**, 257–270. C8[75]

Cayley, A., 1858. "Note on the expansion of the true anomaly," *Quarterly Journal of Mathematics*, **2**, 229–232. C3[34]

Cayley, A., 1861. "Tables of the developments of functions in the theory of elliptic motion," *Memoires of the Royal Astronomical Society*, **29**, 191–306. I

Cayley, A., 1862. "On Lambert's Theorem for Elliptic Motion," *Monthly Notices of the Royal Astronomical Society*, **22**, 238–242.

Charlier, C.L., 1927. *Die Mechanik des Himmels,* Volume II, de Gruyter, Berlin, 255–277. C2

Cherry, T.M., 1955. "On Kepler's equation," *Proceedings of the Cambridge Philosophical Society*, **51**, 81–91. F1

Christom, 1744. *Systema Keplerianum,* London. H2

Colwell, P., 1991."Kepler's equation and Newton's method," *Celestial Mechanics*, **52**, 203–204. D2

Colwell, P., 1992. "Bessel functions and Kepler's equation," *The American Mathematical Monthly*, **99**, 45–48. H2

Conway, B.A., 1986. "An improved algorithm due to Laguerre for the solution of Kepler's equation," *Celestial Mechanics*, **39**, 199–211. F2K[104]

Creedy, C., 1857. "General and practical solution of Kepler's problem," *Quarterly Journal of Pure and Applied Mathematics*, **1**, 259–272. C2[25]

Cronstrand, S.A., 1813. "Bestimmung der Zeit, wenn die Anomalie gegeben und die Bahn nicht viel von der Parabel abweicht," *Handlingar av Svenska Vetenskapsakademien*, 13–21. G

D'Amario, L.A. and Synnott, S.P., 1969. "A comparison of solutions of Kepler's and Lambert's problems," *MIT Instrumentation Laboratory Report T-520.* H5

D'Ocagne, M., 1894. "Abaque en points isoplethes de l'équation de Kepler," *Bulletin des Sciences Mathématiques*, **22**, 197–204. J5[91]

Danby, J., 1987. "The solution of Kepler's equation, III," *Celestial Mechanics*, **40**, 303–312. F2K[104]

Danby, J., 1988. *Fundamentals of Celestial Mechanics,* Second revised and enlarged edition, Willmann-Bell, Richmond. H5

Danby, J. and Burkhardt, T., 1983. "The solution of Kepler's equation, I," *Celestial Mechanics*, **31**, 95–107. H2[102]

Dass, R., 1990. "Minimum modulus solution of an equation," M.S. paper, Department of Mathematics, Iowa State University. F1[95]

Davidson, M., 1933. "Determination of the true anomaly and the time of perihelion passage in hyperbolic orbits," *Monthly Notices of the Royal Astronomical Society*, **93**, 33–39. G

Davidson, M., 1934. "Note on Dr. Smiley's paper," *Monthly Notices of the Royal Astronomical Society*, **95**, 66–67. G

Dazew, A., "Ein mechanische Weg zur Losung der Keplerschen Gleichung," *Astronomische Nachrichten*, **253**, 191–192. J2[88]

De Gasparis, A., 1856. "Sul metodo del Prof. Scorza per risolvere il problema di Keplero," *Rendiconto della adunanze e de'lavori della Reale Accademia della Scienze (Fiz. e Mat.) di Napoli,* **5,** 49–61. F2

De Gasparis, A., 1857a. "Table pour la solution du problème de Kepler," *Astronomische Nachrichten,* **46,** 17–64. J3[81]

De Gasparis, A., 1857b. "Sur une méthode expéditive pour obtenir la valeur de l'anomalie excentrique," *Comptes Rendus (Paris),* **44,** 338–339. J3[81]

De Gasparis, A., 1857c. "Formole e tavole per la soluzione del problema di Keplero," *Il Giambattista-Vico, giornale scientifico, Napoli,* **1,** 113–120. J3[81]

De Gasparis, A., 1857d. "Formole e tavole per la soluzione del problema di Keplero," *Memorie della Reale Accademia della Scienze (Napoli),* **2,** 151–155. J3[81]

De Gasparis, A., 1857e. "Nota alla soluzione del problema di Keplero data dal Prof. N. Fergola," *Memorie della Reale Accademia della Scienze (Napoli),* **2,** 173–174.

De Gasparis, A., 1857f. "On a numerical solution of Kepler's Problem reduced to Tabular Form," *Monthly Notices of the Royal Astronomical Society,* **17,** 155–158. J3[81]

De Gasparis, A., 1858. "Schreiben des Herrn de Gasparis an der Herausgeber," *Astronomische Nachrichten,* **47,** 8ff. J3

De Gasparis, A., 1862a. "Schreiben Herrn Annibal de Gasparis an der Herausgeber (Règle pour résoudre le problème de Kepler)," *Astronomische Nachrichten,* **57,** 333–336. J3

De Gasparis, A., 1862b. "Règle pour la solution du problème de Kepler," *Comptes Rendus (Paris),* **54,** 1195–1196. J3

De Gasparis, A., 1862c. "Regola per la soluzione del problema di Keplero," *Rendiconto della adunanze e de'lavori della Reale Accademia della Scienze,* **1,** 131–134. J3

De Gasparis, A., 1863. "Rule for the solution of Kepler's problem," *Monthly Notices of the Royal Astronomical Society,* **23,** 20–22. J3

De Gasparis, A., 1877a. "Altra soluzione numerica del problema detto di Keplero," *Atti della Reale Accademia della Scienze, Fisiche e Matematiche, Napoli,* **16,** 17–21. J3

De Gasparis, A., 1877b. "Sur le problème de Kepler," *Comptes Rendus (Paris),* **84,** 333–335. J3

De Gasparis, A., 1877c. "On Kepler's problem," *Monthly Notices of the Royal Astronomical Society,* **37,** 263–265. J3

De Gasparis, A., 1879a. "Sopra alcuni elementi ellittici in funzione dell'anomalia media espressa in parti del raggio," *Atti della Reale Accademia dei Lincei (3), Transunti, Rome,* **3,** 111–112. C7[43]

De Gasparis, A., 1879b. "On some formulae for expressing the value of the eccentric anomaly etc. in terms of the mean anomaly," *Monthly Notices of the Royal Astronomical Society,* **39,** 386–387.

De Gasparis, A., 1881a. "Saggio di una tavola numerica per la soluzione del problema di Keplero," *Rendiconto della Reale Accademia della Scienze (Fiz. e Mat.) di Napoli,* **20,** 177–179. I

De Gasparis, A., 1881b. "Tavola per la soluzione numerica del problema di Keplero," *Rendiconto della Reale Accademia della Scienze (Fiz. e Mat.) di Napoli,* **20,** 234–238. J3

De Gasparis, A., 1881c. "Altre serie fra anomalie e raggio vettore nelle ellissi planetaire," *Rendiconto della Reale Accademia della Scienze (Fiz. e Mat.) di Napoli,* **20,** 260–264. D1

De Gasparis, A., 1882. "Sur le problème de Kepler," *Comptes Rendus (Paris),* **95,** 446. C

De la Caille, N.L., 1750a. "Sur les éléments de la théorie du soleil, Premier Mémoire," *Mémoires de l'Académie Royale des Sciences (Paris),* 11–27. A2[46]

De la Caille, N.L., 1750b. "Sur les éléments de la théorie du soleil, Second Mémoire," *Mémoires de l'Académie Royale des Sciences (Paris),* 166–178. A2

De la Hire, P., 1713. "Remarques sur le mouvement des planètes, et principalement sur celui de la Lune," *Mémoires de l'Académie Royale des Sciences (Amsterdam),* 392–404. F2

Degen, C.F., 1821. "Ueber das Kepler'che Problem," *Astronomisches Jahrbuch,* 88–99. C2[25]

Degen, C.F., 1823. "Approximatorische Bestimmung der wahren Anomalie aus der mittleren nach einer (veilleicht) nicht sehr bekannter Methode," *Astronomische Nachrichten,* **1,** 229–230. C2

Delambre, J.-B.J., 1812. "Review of Gauss' Theoria motus," *Connaissance des Temps.* H6

Delambre, J.-B.J., 1820. "Série de A. Robertson pour calculer l'anomalie excentrique étendue à 5 heures de plus par Delambre," *Connaissance des Temps.* C

Delves, L. and Lyness, J., 1967. "A numerical method for locating the zeros of an analytic function," *Mathematics of Computation,* **21,** 543–577. H4[112]

Detmoldt, W.H., 1798. "Kepleri problema celebri," *Commentationes Societatis Regiae Gottingensis.* H2

Doberck, W., 1878. "Values of M tabulated for different e with argument E," *Astronomische Nachrichten,* **92,** 275–282. I[82]

Dolainski, F., 1913. "Graphisches Verfahren zur Losung der Keplerschen Gleichung," *Astronomische Nachrichten,* **194,** 43–44. B1[21]

Draper, G.H., 1936. "Table for the solution of Kepler's equation," *Astronomical Journal,* **45,** 140–143. I[85]

Dubois, E., 1863. "Moyen de résoudre graphiquement le problème de Kepler," *Astronomische Nachrichten,* **59,** 177–180. B1[21]

Duffet-Smith, P., 1981. *Practical Astronomy with Your Calculator,* Cambridge University Press, Cambridge, 120–129. H5J5[89]

Ellicott, A., 1799. "A method of calculating the eccentric anomaly of the planets," *Transactions of the American Philosophical Society,* **4,** 67–69. A2K[14]

Encke, J.F., 1850. "Ueber die Auflosung der Kepler'schen Gleichung," *Astronomische Nachrichten,* **30,** 277–292. A2K[14]

Estes, R.H. and Lancaster, E.R., 1970. "A power series inverse in time of the universal Kepler equation," *Journal of the Astronautical Sciences,* **18,** 168–172. E

Euler, L., 1740. "De motu planetarum et orbitarum determinatione," *Commentarii Acad. Sci. Petrop.,* **7,** 67–85. D1[19]

Euler, L., 1748. "Mémoire sur la plus grande équation des planètes," *Mémoires de l'Académie Royale des Sciences (Berlin)* **2,** 225–248. D1

Euler, L., 1750. "Emendatis tabularum astronomicarum per loca planetarum geocentrica," *Commentarii Acad. Sci. Petrop.,* **12,** 109–221. D1

Fill, T.J., 1976. "Extension of Gauss' method for the solution of Kepler's equation," M.S. Thesis, MIT. D4

Flauti, V., 1857. "Il problema di Keplero risoluto dal fu illustre geometra N. Fergola, estratto da'suoi MSS," *Memorie della Reale Accademia della Scienze (Napoli),* **2,** 156–164. A

Fleckenstein, J.O., 1941. "Notiz zur Lagrangeschen Losung des Keplerschen Problems," *Commentarii Mathematici Helvetici,* **13,** 83–89. C2[37]

Fry, T.C., 1916. "Graphical solution of the position of a body in an elliptic orbit," *Astronomical Journal,* **29,** 141–146. J5[91]

Fuss, N., 1788. "Remarque sur une nouvelle méthode de trouver l'anomalie excentrique par l'anomalie moyenne," *Nova Acta Acad. Sci. Petrop.,* **3,** 302–306. D1

Gauss, C.F., 1906. "Nachlass und Veroffentlichung II. Zur Berechnung der wahren Anomalie in der parabolischen Bewegung," in *Werke,* Band 7, konig. Gessell. Wiss. (Gottingen), Teubner, Leipzig, 351–371.

Gauss, C.F., 1809. *Theoria motus corporum coelestium,* 1963 English translation by C.H. Davis, Dover, New York. D4[64]

Gaythorpe, S.B., 1920. "An approximate solution of Kepler's equation," *Journal of the British Astronomical Association,* **30,** 94–97. A1[17]

Gaythorpe, S.B., 1925. "On Horrocks' treatment of the evection and the equation of the centre with a note on the elliptic hypothesis of Albert

Curtz and its correction by Boulliau and Newton," *Monthly Notices of the Royal Astronomical Society,* **85,** 858–865. A1[8]

Gaythorpe, S.B., 1957. "Jeremiah Horrocks and his 'New Theory of the Moon,'" *Journal of the British Astronomical Association,* **67,** 134–144. A1[10]

Gedeon, G.S., 1964. "Inversion of Kepler's equation and Herrick's unified and universal time of flight equations," *Journal of the Astronautical Sciences,* **11,** 79–84. E[28]

Glaisher, J.W.L., 1877a. "On an elliptic function solution of Kepler's problem," *Monthly Notices of the Royal Astronomical Society,* **37,** 366–386. F3[35]

Glaisher, J.W.L., 1877b. "On the solution of Kepler's problem," *Monthly Notices of the Royal Astronomical Society,* **37,** 445–458. C2

Gonggrijp, B., 1901. "Ueber eine graphometrische Losung der Kepler'schen Gleichung (und die Construction der heliocentrischen Coordinaten eines Planeten)," *Astronomische Nachrichten,* **155,** 369–374. B2[21]

Gooding, R.H., 1987a. "Solution of the hyperbolic Kepler's equation," *Royal Aircraft Establishment Technical Report* 87042. K[104]

Gooding, R.H., 1987b. "Universal procedures for conversion of orbital elements to and from position and velocity (unperturbed orbits)," *Royal Aircraft Establishment Technical Report* 87043. E[104]

Gooding, R.H., 1988. "On the solution of Lambert's orbital boundary-value problem," *Royal Aircraft Establishment Technical Report* 88027.

Gooding, R.H., 1990. "A procedure for the solution of Lambert's orbital boundary-value problem," *Celestial Mechanics,* **48,** 145–165.

Gooding, R.H. and Odell, A.W., 1985. "A monograph on Kepler's equation," *Royal Aircraft Establishment Technical Report* 85080. H2[102]

Gooding, R.H. and Odell, A.W., 1988. "The hyperbolic Kepler equation (and the elliptic equation revisited)," *Celestial Mechanics,* **44,** 267–282. K

Gourier, G., 1878. "Sur l'équation de Kepler," *Annales sci. de l'Ecole Normal Superieure,* **7,** 73–76. F1

Grabiner, J., 1981. *The Origins of Cauchy's Rigorous Calculus,* MIT Press, Cambridge. H3[67]

Grandjean de Fouchy, J.P., 1738. "Pour déterminer par observation l'excentricité de la terre, et celle des planètes inférieures," *Mémoires de l'Académie Royale des Sciences (Paris),* 185–192. F2

Grandjean de Fouchy, J.P., 1740. "Second Mémoire sur l'excentricité des planètes," *Mémoires de l'Académie Royale des Sciences (Paris),* 235–242. F2

Grattan-Guiness, I. (with J.R. Ravetz), 1972. *Joseph Fourier, 1768–1830,* MIT Press, Cambridge. H4[27]

Greathead, S.S., 1837. "Investigation of the general term of the expansion of the true anomaly in terms of the mean," *Cambridge Mathematical Journal*, **1**, 208–212. C3[34]

Grunert, J.A., 1851. "Ueber das Kepler'sche Problem," *Astronomische Nachrichten*, **33**, 209–218. D1[47]

Grunert, J.A., 1856. "Neue naherungsweise Auflosung der Keplerschen Aufgabe," *Sitzungsbericht Akad. Wien*, **19**, 3–9. A1[17]

Gylden, H., 1875a. *En ny losning till det Keplerska problemet*, Forhandlingar Stockholm. F3[35]

Gylden, H., 1875b. "Zur Auflosung des Kepler'schen Problemes," *Vierteljahrschrift der Astronomischen Gesellschaft*, **10**, 285–296. F3[35]

Hagihara, Y., 1936. "An application of Cauchy's theorem of residues for functions of two variables to the expansions in planetary theory," *Astronomical Journal*, **45**, 129–132. F1[37]

Hall, A., 1905. "Elliptic motion," *Popular Astronomy*, **13**, 287–296. C3

Halley, E., 1694. "A new and general method for finding the roots of equations," *Philosophical Transactions of the Royal Society*, **18**, 136–148. H4

Halley, E., 1705. "Astronomiae cometicae synopsis," *Philosophical Transactions of the Royal Society*, **24**, 1882–1899. G1[101]

Hansen, P.A., 1852a. "Ueber die Berechnung der Storungen durch mechanische Quadraturen," *Astronomische Nachrichten*, **34**, 101–142. C3

Hansen, P.A., 1852b. "Neue directe Auflosung des Keplerschen Problems," *Berichte uber die Verhandligen d. konig. sachsischen Gess.d. Wiss. zu Leipzig, Math-Phys. Cl.*, **4**, 55–56. C3[33]

Hansen, P.A., 1852c. "Nouvelle solution du problème de Kepler," *Comptes Rendus (Paris)*, **35**, 746–747. C3[33]

Hansen, P.A., 1853a. "Neue directe Auflosung des Kepler'schen Problems," *Astronomische Nachrichten*, **35**, 317–322. C3[33]

Hansen, P.A., 1853b. "Uber die Entwickelung des Products einer Potenz des Radius Vectors mit dem Sinus oder Cosinus eines Vilefachen der wahren Anomalie in Reihen, die nach der Sinussen oder Cosinussen der Vielfachen der wahren, excentrischen oder mittleren Anomalie fortschriften," *Berichte uber die Verhandligen d. konig. sachsischen Gess. d. Wiss. zu Leipzig, Math-Phys. Cl.*, **5**, 1–14. C3

Hansen, P.A., 1853c. "New solution of Kepler's problem," *Monthly Notices of the Royal Astronomical Society*, **13**, 52. C3[33]

Hansen, P.A., 1855. "Entwickelung des Products einer Potenz des Radius Vectors mit dem Sinus oder Cosinus eines Vielfachen den wahren Anomalie in Reihen, die nach den Sinussen oder Cosinussen der Vielfachen der wahren, excentrischen oder mittleren Anomalie fortschreiten,"

Abhandlungen d. Math-Phys. Cl. d. konig. sachsischen Gess. d. Wiss. zu Leipzig, **2**, 181–281. C3[33]

Harris, S., 1988. "Johannes Kepler's uphill battle," *American Scientist,* **76**, 575.

Hartmann, J., 1918. "Zur Auflosung der Keplerschen Gleichung," *Astronomische Nachrichten,* **205**, 309–312. F2[86]

Hauff, J., 1797. "Uber die Anomalien," *Astronomisches Jahrbuch,* 123–134. H5

Hennert, J.F., 1804. "Aud der grosse Mittelspunkts-Gleichung die Excentricitat zu finden," *Astronomisches Jahrbuch,* 218–226. C3[31]

Henrici, P., 1974. *Applied and Computational Complex Analysis,* Vol. I, Wiley-Interscience, New York. H4

Henrici, P., 1986. *Applied and Computational Complex Analysis,* Vol. III, Wiley-Interscience, New York. H4[111]

Herget, P., 1936. "Determination of orbits, with tables for true anomaly and perihelion passage," *Publications of the Cincinnati Observatory,* 1–11. G1I

Herglotz, G., 1910. "Bahnbestimmung der Planeten und Kometen," *Enzyklopedia d. Math. Wiss., (2),* **6**, 379–426. H1

Hermann, J., 1726. "Geminus modus directus divideni semicirculum in data ratione: quibus Keplerianum problema de inveniendis planetarum locis ad datum quodvis tempus, solutum exhibetur," *Comm. Acad. Sci. Imp. Petrop.,* **1**, 142–148. A2K[14]

Herrick, S., 1945. "'Nearly parabolic' and 'nearly rectilinear' orbits," *Astronomical Journal,* **51**, 123. G[94]

Herrick, S., 1971. *Astrodynamics,* Van Nostrand Reinhold, New York. H5[96]

Herz, N., 1881. "Note zur Losung des Kepler'schen Problem," *Astronomische Nachrichten,* **99**, 31–32. A2I[83]

Houzeau, J.C. and Lancaster, A., 1882. *Bibliographie générale de l'astronomie,* **2**, 456–461. H1

Howe, H.A., 1879. "Three approximate solutions of Kepler's problem," *Journal of the Cincinnati Society of Natural History,* **2**, 205–210. A1[17]

Howe, H.A., 1880a. "A new approximate solution of Kepler's problem," *Astronomische Nachrichten,* **97**, 273–276. A1[17]

Howe, H.A., 1880b. "Two new solutions of Kepler's problem," *Astronomische Nachrichten,* **98**, 305–308. A1[17]

Howe, H.A., 1884. "A short method for Kepler's problem," *Astronomische Nachrichten,* **108**, 427–430. A1[17]

Howe, H.A., 1919. "A desideratum in solving Kepler's problem," *Popular Astronomy,* **27**, 569. A1[17]

Howe, H.A., 1922. "Kepler's problem for high planetary eccentricities," *Popular Astronomy,* **30**, 161. A1K

Hurwitz, A., 1906. "Sur les points critiques des fonctions inverses," *Comptes Rendus (Paris)*, **143,** 877–879. H4[76]

Innes, R.T.A., 1924. "Elliptic expansions," *Astronomical Journal*, **36,** 39–40. C3

Innes, R.T.A., 1932. "The parabolic orbit," *Astronomische Nachrichten*, **246,** 431–432. G1[94]

Innes, R.T.A., 1933. "To find the eccentric anomaly," *Observatory*, **56,** 61–62. I[94]

Ioakimidis, N.I. and Papadakis, K., 1985. "A new simple method for the analytical solution of Kepler's equation," *Celestial Mechanics*, **35,** 305–316. F1M2[111]

Iushkevich, A.P., 1966. *Geschichte der Mathematik in Mittelalter*, Pfalz Verlag, Basel, 324–325. H3[5]

Iushkevich, A.P., 1976. "Les Mathématiques Arabes (VIIIe-XVe siècles)," *Collection d'Histoires des Sciences, 2-C.N.R.S.*, Paris, 162–163. H3[5]

Ivory, J., 1805. "A new and universal solution of Kepler's problem," *Transactions of the Royal Society of Edinburgh*, **5,** 203–246. D3[54]

Jackson, A., 1988. "The AMS centennial, 1888–1988," *Notices of the American Mathematical Society*, **35,** 15–20. H3[36]

Jacobi, C.G.J., 1836. "Formula transformationis integralium definitorum," *Journal für die reine und angewandte Mathematik*, **15,** 1–38. C3

Jacobi, C.G.J., 1848. "Ueber die annaherunde Bestimmung sehr entfernter Gleiden in der Entwickelung der elliptischen Coordinaten nebst einer Ausdehnung der Laplaceschen Method zur Bestimmung der Functionen gerader Zahlen," *Astronomische Nachrichten*, **28,** 257–270. C3

Jacobi, C.G.J., 1850. "Untersuchungen uber die Convergenz der Reihe durch welche das Kepler'sche Problem gelost wird," *Astronomische Nachrichten*, **30,** 197–254. C3[33]

Jekhowsky, M.B., 1918. "Application des fonctions de Bessel à plusieurs variables au mouvement elliptique," *Bulletin Astronomique*, **35,** 139–145. C4[36]

Jensen, J., Townsend, G., Kork, J. and Kraft, D., 1962. *Design Guide to Orbital Flight*, Chapter II, McGraw-Hill, New York. B3[91]

Jorgensen, N.T., 1974. "Peter Horrebow's solution of the Kepler-problem," *Centaurus*, **18,** 173–183. A1K[9]

Kaminskii, B.I., 1982. "Expansion of the coordinates of elliptical motion with respect to the Levi-Civita parameter," *Soviet Astronomy*, **26,** 478–480. C6[38]

Kapteyn, J.C., 1883. "Ueber das Kepler'sche Problem," *Copernicus*, **3,** 25–34. C1K

Kapteyn, M.W., 1893. "Recherches sur les fonctions de Fourier-Bessel," *Ann. Sci. de l'Ecole Normale Superieure (3)*, **10**, 91–122. C3[39]

Karlinski, F., 1862a. "Uber das Kepler'sche Problem speciell bei der Planeten," *Astronomische Nachrichten*, **57**, 183–190. D1K

Karlinski, F., 1862b. "Bemerkungen uber die Anwendung der Gauss' ischen Methode der Auflosung des Kepler'chen Problems bei grossen Excentricitaten," *Astronomische Nachrichten*, **57**, 191–192. D4

Karlinski, F., 1862c. "Ueber die schnellste Praxis der Auflosung der Keplerschen Gleichung, $M = E - e \sin E$, E bei grossen Excentricitaten der elliptischen Cometenbahnen," *Sitzungsber. d. konig. Bohmischen-Gessell. der Wiss. in Prag*, 41–43. H2

Keill, J., 1713. "Problematis Kepleriano, de inveniendo vero motu planetarum, areas tempori proportionales in orbibus ellipticus circa focorum alterum describentium, solutio Newtoniana," *Philosophical Transactions of the Royal Society*, **28**, 1–10. A3[51]

Kennedy, E.S., 1956. "Parallax theory in Islamic astronomy," *Isis*, **47**, 33–53. H3[5]

Kennedy, E.S. and Transue, W.R., 1956. "A medieval iterative algorism," *The American Mathematical Monthly*, **63**, 80–83. H3[5]

Kletke, C.A., 1852. *Directe Auflosung des Kepler'schen Problems*, Breslau. I

Kline, M., 1972. *Mathematical Thought from Ancient to Modern Times*, Oxford University Press, Oxford. H3[20]

Kline, M., 1983. "Euler and infinite series," *Mathematics Magazine*, **56**, 307–314. H4[23]

Klugel, G., 1789. "Formel, aus der Anomalia excentri die mittlere Anomalie zu finden," *Astronomisches Jahrbuch*, 191–193. C2[25]

Klugel, G., 1806. "Neue Art, die grosste Mittelpunckts-Gleichung aur der Excentricitat zu finden," *Astronomisches Jahrbuch*, 238–241. F2

Klumpp, A., 1990. "Universal Lambert and Kepler algorithms for autonomous rendevous," *AIAA Astrodynamics Conference*, AIAA-90–2883. K

Klumpp, A., 1991. "Performance comparison of Lambert and Kepler algorithms," Jet Propulsion Laboratory Interoffice Memorandum. K[107]

Knobel, E., 1876. "Reference catalogue of astronomical papers and researches," *Monthly Notices of the Royal Astronomical Society*, **36**, 365–392. H1

Kreutz, H., 1880. "Review of Oppolzer's 'Ueber die Berechnung der wahren Anomalie in nahezu parabolischen Bahnen,'" *Vierteljahrschrift der Astronomische Gesellschaft*, **15**, 219–222. H6

Kung, H. and Traub, J., 1978. "All algebraic functions can be computed fast," *Journal of the Association for Computing Machinery*, **25**, 245–260. H4[156]

Lagrange, J.L., 1770. "Nouvelle méthode pour résoudre les équations littérales pour le moyen des séries," *Mémoires de l'Académie Royale des Sciences (Berlin)*, **24**. C2[24]

Lagrange, J.L., 1771. "Sur le problème de Kepler," *Mémoires de l'Académie Royale des Sciences (Berlin)*, **25**, 204–233. C2[24]

Lalande, J.J., 1796. "Erreur de Kepler sur la courbe de l'orbite apparente de la lune," *Connaissance des Temps*, **6**.

Langebartel, R.G., 1948. "A note on Kepler's equation," *Astronomical Journal*, **53**, 101–104. F1[94]

Laplace, P.S., 1823. "Mémoire sur le développement de l'anomalie vraie et du rayon-vecteur elliptique, en séries ordonnées suivant les puissances de l'excentricité," *Mémoires de l'Académie Royale des Sciences (Paris)*, **6**, 61–80. C8[26]

Laplace, P.S., 1828. "Sur le développement en séries du radical qui exprime la distance mutuelle de deux planètes, et du rayon vecteur elliptique," *Connaissance des Temps*.

Laplace, P.S., 1966. *Celestial Mechanics Reprint* of the 1839 translation and Commentary by Nathaniel Bowditch of *Traité de mécanique céleste*, Chelsea Publishing, Bronx, N.Y. C2[59]

Laska, V., 1914. "Ein Nomogram zur Auflosung der Kepler'schen Gleichung," *Astronomische Nachrichten*, **199**, 295–296. J5[91]

Laska, W., 1884. "Note zur Losung der Kepler'schen Problems," *Astronomische Nachrichten*, **111**, 9–10. C3

Le Grand Roy, E., 1903. "Résolution graphique de l'équation de Kepler, d'après Radau," *Archives des sciences physiques et naturelles*, **16**, 328. J5[91]

LeFort, F., 1846. "Expressions numérique des intégrales définies qui se présentent quand on cherche les termes généraux du développement des coordonnées d'une planète, dans son mouvement elliptique," *Journal de Mathématiques*, **11**, 142–152. C3

Lehmann, W., 1855. "Aus einem Schreiben des Herrn W. Lehmann an des Herausgeber," *Astronomische Nachrichten*, **39**, 257–260. H5[80]

Lehmann, W., 1856. "Construction einer neuen Tafel fur den lapsus ellipticus innerhalb der letzten funf Achtel des Zeitsraums, in welche der bewegte Punct aus dem Zustande der Ruhe bis zum Centralpunct gelangt," *Astronomische Nachrichten*, **44**, 161–170. H5

Lehmann, W., 1857. "Schreiben des Herrn Dr. Lehmann an den Herausgeber," *Astronomische Nachrichten*, **46**, 257–260. I

LeVerrier, U.J.J., 1855. "Recherches Astronomiques," *Annales de l'Observatoire de Paris*, **1**, 182–231. C

Levi-Civita, T., 1904a. "Sopra la equazione di Kepler," *Astronomische Nachrichten*, **164**, 313–314. C6[38]

Levi-Civita, T., 1904b. "Sopra la equazione di Kepler," *Rendiconti della Reale Accademia dei Lincei, Classe di scienze fisiche, matematiche e naturali (5)*, **13**, 260–268. C6[38]

Levin, A.E., 1922. "The slide-rule as an auxiliary to the solution of Kepler's equation," *Monthly Notices of the Royal Astronomical Society*, **82**, 177–180. J3[88]

Lindqvist, J.H., 1778. "Nytt satt at til en gifven Anomalia media finna en Planets Anomalia vera," *Kongl. Svenska Vetenskaps Akademiens Handlingar (Stockholm)*, **39**, 136–141. C

Mache, H., 1910. "Uber die Anwendung des logarithmischen Rechen-schreibers zur Losung der Keplerschen Gleichung," *Astronomische Nachrichten*, **186**, 357–358. J3[88]

Machin, J., 1738. "Solution of Kepler's problem," *Philosophical Transactions of the Royal Society*, 205–230. F2K[52]

Marth, A., 1890a. "On the computation of the equation of the centre in elliptical orbits of moderate eccentricities," *Monthly Notices of the Royal Astronomical Society*, **50**, 502–510. A1[18]

Marth, A., 1890b. "A simple solution of Kepler's problem," *Monthly Notices of the Royal Astronomical Society*, **50**, 511. D1I[18]

Marth, A., 1890c. "Two auxiliary tables for the solution of Kepler's problem," *Monthly Notices of the Royal Astronomical Society*, **50**, 530–546. A1I[83]

Meeus, J., 1991. *Astronomical Algorithms*, Willmann-Bell, Richmond.

Melander, D., 1767. "Expositio inventorum Kepleri," *Acta Societatis Regiae Upsaliensis*. H2

Mercator, N., 1664. *Hypothesis astronomica nova, et consensus eius cum observationibus*. A1[11]

Mercator, N., 1670. "Some considerations of Mr. Nic Mercator, concerning the geometrik and direct method of Signior Cassini for finding the apogees, ecentricities and anomalies of the planets," *Philosophical Transactions of the Royal Society*, **5**, 1168–1175. A2[11]

Mikkola, S., 1987. "A cubic approximation for Kepler's equation," *Celestial Mechanics*, **40**, 329–334. F2[106]

Moller, J.P., 1933. "Calculation of ephemerides in nearly parabolic orbits," *Monthly Notices of the Royal Astronomical Society*, **93**, 777–788. GI[94]

Moller, J.P., 1940. "On the solution of Kepler's equation," in *Festschrift fur Elis Stromgren*, Munksgaard, Copenhagen, 163–174. GI[94]

Moore, P., 1987. *Astronomers' Stars*, Norton, New York. H3[58]

Morris, W. (editor), 1981. *The American Heritage Dictionary of the English Language*, Houghton Mifflin, Boston.

Morrison, J., 1883. "On the computation of the eccentric anomaly, equation of the centre, and radius vector of a planet, in terms of the mean anomaly and eccentricity," *Monthly Notices of the Royal Astronomical Society,* **43,** 345–365,494. C2

Mossotti, O.F., 1817a. "Nuova analisi del problema di determinare le orbite dei corpi celeste," *Effemeridi astronomiche di Milano,* Part I, Appendice, 33–112. H2

Mossotti, O.F., 1817b. "Neue Auflosung des Problems die Bahn eines Himmelkorpers zu bestimmen," *Zeitschrifte für Astronomie,* **3.**

Moulton, F.R., 1895. "An important method of solving Kepler's equation," *Popular Astronomy,* **3,** 136–141. B1J4[21]

Moulton, F.R., 1903. "The true radii of convergence of the expressions for the ratios of the triangles when developed as power series in the time-intervals," *Astronomical Journal,* **23,** 93–102. C7[43]

Moulton, F.R.,1907. "Direct computation of the expressions for the coordinates in elliptic motion," *Astronomical Journal,* **25,** 145–149. C2[26]

Moulton, F.R., 1914. *An Introduction to Celestial Mechanics,* Second Edition, McGraw-Hill, New York, 161ff. H5[26]

Moulton, F.R., 1930. "Metodo grafico para resolver la ecuacion de Kepler," *Revista Astronomica,* **2,** 409–413. B1J4[94]

Neutsch, W. and Schrufer, E., 1986. "Simple integrals for solving of Kepler's equation," *Astrophysics and Space Science,* **125,** 77–83. F1M2

Newton, I., 1934. *Sir Isaac Newton's Mathematical Principles of Natural Philosophy and his System of the World,* translated into English by Andrew Motte in 1729. The translations revised, and supplied with a historical and explanatory appendix by Florian Cajori, University of California Press, Berkeley, 112–116, 647–648. D2[48]

Ng, E., 1979. "A general algorithm for the solution of Kepler's equation for elliptic orbits," *Celestial Mechanics,* **20,** 243–249. D2K[99]

Nicolai, B., 1844. "Schreiben des Herrn Hofraths Nicolai an den Herausgeber," *Astronomische Nachrichten,* **21,** 77–80. G1I

Nicollic, 1746. "Sur la détermination des orbites planétaires, où l'on demontre quelques nouvelles propriétés des sections coniques," *Mémoires de l'Académie Royale des Sciences (Paris),* 291–318. D2

Nijenhuis, A., 1991. "Solving Kepler's equation with high efficiency and accuracy," *Celestial Mechanics,* **51,** 319–330. F2K[107]

Nordmark, Z., 1789. "Om integralen af den foranderliga sectorn i det fa kallade Kepler's problem," *Kongl. Svenska Vetenskaps Akademiens Handlingar,* **10,** 225–232. H5

Nordmark, Z., 1799. "Tentamen seriei apte convergentis in solutione Problematis Kepleri," *Acta Societatis Regiae Upsaliensis,* **6,** 246–269. H2

Odell, A.W. and Gooding, R.H., 1986. "Procedures for solving Kepler's equation," *Celestial Mechanics,* **38,** 307–334. H2[102]

Olbers, W., 1797. *Abhandlung uber die leichteste und bequemste Methode die Bahn eines Cometen aus einigen Beobachtungen zu berechnen,* Wiemar. G1

Oppolzer, T., 1878. "Entwicklung der Differentialquotienten der wahren Anomalie und des Radius-vector nach der Excentricitat in nahezu parabolischen Bahnen," *Monatsberichte der kongliche Preussischen Akademie der Wissenschaften zu Berlin,* 583–602. C3

Oppolzer, T., 1879. "Entwicklung der Differentialquotienten der wahren Anomalie und des Radiusvector nach der Excentricitat in nahezu parabolischen Bahnen," *Astronomische Nachrichten,* **95,** 13–16. GI[66]

Oppolzer, T., 1880. "Ueber die Berechnung der wahren Anomalie in nahezu parabolischen Bahnen," *Abhandlungen der konigl. bayer. Akad. d. Wiss., Math-Phys. Cl.,* **13,** 137–168. GI[66]

Oppolzer, T., 1882. *Lehrbuch zur Bahnbestimmung der Kometen und Planeten,* Verlag von W. Engelmann, Leipzig. H5[83]

Oppolzer, T., 1885. "Uber die Auflosung des Kepler'schen Problems," *Denkschriften der kaiserlichen Akademie der Wissenschaften (Wien), Math-Nat. Cl.,* **50,** 185–243. A1I[18]

Osgood, W.F., 1905. "Allgemeine Theorie der analytischer Funktionen a) ciner und b) mehrerer komplex Grossen," in *Encyklopadie der mathematischen Wissenschaften mit Enschluss ihrer Anwendungen,* Volume 2, H. Burkhardt, W. Wirtingen, R. Fricke, E. Hilb (eds.), Teubner, Leipzig, 1–114. H4[76]

Pacassi, J.B., 1794. Extrait de quelques lettres de M. le Baron de Pacassi, Correspondant de l'Académie à Vienne, adressées dans le cours de cette année à M. l'Académicien Fuss (Traduit de l'Allemand), *Nova Acta Acad. Sci. Petrop.,* **8,** 40–41. D1

Pagels, H.R., 1985. *Perfect Symmetry,* Simon and Schuster, New York. H3[30]

Paul, E.W., 1963. "Validity of series expansions of Kepler's equations," *AIAA Journal,* **7,** 1659–1660. C7[43]

Perrey, A., 1853. "Sur la détermination de l'orbite des planètes et des comètes (d'après Cauchy)," *Connaissance des Temps.* H5

Peters, C.F.W., 1891a. "Eine Bemerkung zum Kepler'schen Problem," *Astronomische Nachrichten,* **126,** 291–292. D1[47]

Peters, C.F.W., 1891b. "Remarque sur le problème de Kepler," *Bulletin Astronomique,* **8,** 144. H6[47]

Peters, R.D., 1984. "Rapidly converging series approximation to Kepler's equation," in *Astrodynamics 1983, Advances in Astronautical Sciences,* **54,** 1039–1047. F2[102]

Picart, M.L., 1906. "Sur le développement des coordonnées dans le mouvement elliptique avec les notationes de M. Poincaré," *Bulletin Astronomique,* **23,** 353–359. C2

Pitkin, E.T., 1965a. "A regularized approach to universal orbit variables," *AIAA Journal,* **3,** 1508–1511. E[95]

Pitkin, E.T., 1965b. "On the inversion of the universal Kepler equation," *Journal of the Astronautical Sciences,* **12,** 149–152. E

Plummer, H.C., 1896. "A graphical method of solving Kepler's equation," *Monthly Notices of the Royal Astronomical Society,* **56,** 317–320. J2[21]

Plummer, H.C., 1903. "An approximation to the value of $\epsilon - \sin\epsilon$," *Astronomische Nachrichten,* **162,** 299–300. F2

Plummer, H.C., 1906. "On some points connected with the determination of orbits," *Monthly Notices of the Royal Astronomical Society,* **66,** 491–498. F2

Plummer, H.C., 1907. "Note on a mechanical solution of Kepler's equation," *Monthly Notices of the Royal Astronomical Society,* **67,** 67–70. J2[21]

Plummer, H.C., 1919. "On Professor Howe's method of solving Kepler's equation," *Monthly Notices of the Royal Astronomical Society,* **80,** 207–211. A1[17]

Poisson, S.D., 1811. "Traité de mécanique." C3[31]

Poisson, S.D., 1825. "Sur une nouvelle manière d'exprimer les coordonnées des planètes dans le mouvement elliptique," *Connaissance des Temps.* C3

Pollard, H., 1976. "Celestial Mechanics," *Carus Mathematical Monographs, No. 18, Mathematical Association of America.* H5[141]

Posselt, J., 1818. "Ueber die Bestimmung der Zeit aus der Anomalie in sehr excentrischen Bahnen," *Zeitschrifte für Astronomie,* **5,** 161–170. C3

Pritchard, C., 1877. "Two mechanical solutions of Kepler's problem," *Monthly Notices of the Royal Astronomical Society,* **37,** 354–359. J4[21]

Proctor, R.A., 1873. "Graphical method for determining the motion of a body in an elliptic orbit under gravity," *Monthly Notices of the Royal Astronomical Society,* **33,** 386–391. B1[21]

Prussing, J.E., 1977. "Bounds on the solution to Kepler's problem," *Journal of the Astronautical Sciences,* **25,** 123–128. K[100]

Prussing, J.E., 1979. "Bounds on the solutions to a universal Kepler's equation," *Journal of Guidance and Control,* **21,** 440–442. EK[100]

Puiseux, V.A., 1849. "Sur la convergence des séries qui se présentent dans la théorie du mouvement elliptique des planètes," *Journal de Mathématiques pures et appliquées,* **14,** 33–39. C8[74]

Puiseux, V.A., 1860. "Mémoire sur le développement en séries des coordonnées des planètes et de la fonction perturbatrice," *Journal de Mathématiques pures et appliquées (2),* **5,** 65–102. C8

Puiseux, V.A., 1884. "Sur la convergence des séries ordonnées suivant les puissances de l'excentricité qui se présentent dans la théorie du mouvement elliptique," in *Oeuvres de Lagrange,* Volume 12, J.-A. Serret and G. Darboux (eds.), Gauthier-Villars, Paris, 341–346. C2[74]

Radau, R., 1882. "Remarques sur le problème de Kepler," *Comptes Rendus (Paris),* **95,** 274–276. A1[18]

Radau, R., 1883. "Remarques concernant le problème de Kepler," *Copernicus,* **3,** 132–133. J5

Radau, R., 1884. "Solution graphique du problème de Kepler," *Bulletin Astronomique,* **1,** 381–388. J5[90]

Radau, R., 1885. "Calcul de l'anomalie vraie pour une excentricité voisine de l'unité," *Bulletin Astronomique,* **2,** 509–515. H2

Radau, R., 1888. "Review of Astrand's 'Om en auxiliaertabel til loesning af Kepler's Problem,'" *Bulletin Astronomique,* **5,** 358–359. H6

Radau, R., 1894. "Remarques concernant le problème de Kepler," *Bulletin Astronomique,* **11,** 289–291. B1

Radau, R., 1900. "Bibliographie du Problème de Kepler," *Bulletin Astronomique,* **17,** 37–47. H1

Rambaut, A.A., 1890. "A simple method of obtaining an approximate solution of Kepler's equation," *Monthly Notices of the Royal Astronomical Society,* **50,** 301–302. J2[21]

Rambaut, A.A., 1906. "A simple method of obtaining an approximate solution of Kepler's equation," *Monthly Notices of the Royal Astronomical Society,* **66,** 519–521. J2[88]

Rambaut, A.A., 1913. "The solution of Kepler's problem," *Astronomical Journal,* **27,** 182–183. J2[21]

Rand, W.C., 1942. "A graphical method for the approximate solution of Kepler's equation," *Journal of the British Astronomical Association,* **52,** 149–153. J5[91]

Rittenhouse, D., 1799. "To determine the true place of a planet in an elliptical orbit, directly from the mean anomaly, by converging series," *Transactions of the American Philosophical Society,* **4,** 21–25. C7[43]

Robertson, A., 1816. "Direct and expeditious methods of calculating the excentric from the mean anomaly of a planet," *Philosophical Transactions of the Royal Society,* 127–137. A2K[14]

Rohde, J.P., 1806. "Uber das Problem, aus der mittlere Lange einen Planeten, dessen wahre Lange zu finden," *Astronomisches Jahrbuch, Supplement,* **4,** 99–101. C

Rouché, E., 1862. "Mémoire sur la série de Lagrange," *Journal de l'Ecole Polytechnique,* **39,** 193–224. C2[75]

Roy, A.E., 1988. *Orbital Motion,* Third Edition, Adam Hilger, Bristol. H5[30]

Russell, J.L., 1964. "Kepler's laws of planetary motion: 1609–1666," *Brit. J. Hist. Sci.,* **2,** 1–24. H3

Rust, F., 1908. "Proposal of a new method to compute a planet's anomaly and radius vector," *Popular Astronomy,* **16,** 36–45. F2

Sarnecki, A.J., 1988. "'Minimal' orbital dynamics," *Acta Astronautica,* **17,** 881–891. E[96]

Scheibner, W., 1856a. "Ueber die asymptotische Werthe der Coefficienten in der nach der mittleren Anomalie vorgenommen Entwickelungen," *Berichten der kong. sachsische Gesell. d. Wiss. (Leipzig),* **8,** 40–64. C3[33]

Scheibner, W., 1856b. "On the asymptotic values of the coefficients in the development of any power of the radius vector according to the mean anomaly," *Gould's Astronomical Journal,* **4,** 177–182. C3[33]

Scheibner, W., 1880a. "Ueber die asymptotischen Werthe der Coefficienten in der Entwickelung einer beleibigen Potenz der Radiusvectors nach der mittleren Anomalie," *Math. Annalen,* **17,** 531–544. C3

Scheibner, W., 1880b. "Ueber die asymptotischen Werthe der Coefficienten in den nach mittleren Anomalie vorgenommenen Entwickelungen," *Math. Annalen,* **17,** 545–560. C3

Schering, E., 1884. "Zur Losung der Kepler'schen Gleichung," *Astronomische Nachrichten,* **109,** 193–200. A1[18]

Schubert, E., 1854. "A new method of solving Kepler's equation," *Astronomical Journal,* **3,** 37–39. D2[53]

Schubert, F.T., 1818. "A new investigation of Kepler's problem," *Memoirs of the American Academy of Arts and Sciences,* **4,** 1–27. C2[25]

Schubert, F.T., 1820a. "Ueber das Kepler'sche Problem," *Astronomisches Jahrbuch,* 113–149. C2[25]

Schubert, F.T., 1820b. "Berechnung der wahren Anomalie eines Planeten aus der mittleren und der Excentricitat, bis zur 13ten Potenz," *Astronomisches Jahrbuch,* 229–234. C2[25]

Sconzo, P., 1933a. "Sulla risoluzione dell'equazione di Kepler," *Atti della Pontificia Accademia della Scienze ("Nuovi Lincei"),* **86,** 55–58. D2

Sconzo, P., 1933b. "Espressione dell'anomalia vera in funzione dell'anomalia media," *Memoria della Societa Astronomica Italiana,* **7,** 59–66. C3[32]

Sconzo, P., 1934. "Sulla risoluzione dell'equazione di Kepler, II," *Atti della Pontifica Accademia della Scienze ("Nuovi Lincei")*, **87**, 490–493. C3[53]

Scorza, G., 1857. "Metodo per rinvenire l'anomalie eccentrica, data la media," *Memoire della Reale Accademia della Scienze di Napoli*, **2**, 165–172. A

Searle, G.M., 1889. "On the computation of the true anomaly, radius-vector and coordinates of great eccentricity," *Astronomical Journal*, **8**, 153–156. GI[66]

Searle, G.M., 1892. "On the computation of places in eccentric ellipses and hyperbolas," *Astronomical Journal*, **11**, 89–92. GI[66]

See, T.J.J., 1895. "A general method for facilitating the solution of Kepler's equation by mechanical means," *Monthly Notices of the Royal Astronomical Society*, **55**, 425–429. J4[21]

See, T.J.J., 1896. "Note on the discovery of the graphical method for solving Kepler's equation by means of a curve of sines," *Monthly Notices of the Royal Astronomical Society*, **56**, 54–55. B1[21]

Serafin, R.A., 1986. "Bounds on the solution to Kepler's equation," *Celestial Mechanics*, **38**, 111–121. K[106]

Serret, J.-A., 1856. "Note sur la condition de convergence des séries qui se présentent dans la théorie du mouvement elliptique des planètes," *Comptes Rendus (Paris)*, **42**, 1134–1137. C8

Serret, J.-A., 1857. "Note sur l'équation dont depend l'anomalie excentrique dans la théorie du mouvement elliptique des planètes," *Comptes Rendus (Paris)*, **44**, 486–490. C8

Serret, J.-A., 1859. "Note sur l'équation dont depend l'anomalie excentrique et sur les séries qui se présentent dans la théorie du mouvement elliptique des corps célestes," *Annales de l'Observatoire de Paris*, **5**, 337–352. C8[75]

Seydler, A., 1887. "Beitrag zur Losung des Problems von Kepler," *Sitzungsberichte d. konigl. bohmischen Gess. d. Wiss. (Math-Nat.)*, 547–558, 734–758. H2

Seydler, A., 1888. "Zur Losung des Kepler'schen Problems," *Astronomische Nachrichten*, **118**, 261–272. I[82]

Sheela, B.V., 1982. "An empirical initial estimate for the solution of Kepler's equation," *Journal of the Astronautical Sciences*, **30**, 415–419. F2M3[101]

Shepperd, S.W., 1985. "Universal Keplerian state transition matrix," *Celestial Mechanics*, **35**, 129–144. E

Shepperd, S.W., 1987. "Naturally occurring continued fractions in the variation of Kepler's equation," *Celestial Mechanics*, **42**, 91–106. E

Siewert, C.E. and Burniston, E.E., 1972. "An exact analytical solution of Kepler's equation," *Celestial Mechanics,* **6,** 294–304. F1M2[109]

Simpson, T., 1740. "From the mean anomaly of a planet, to find its place in orbit," in *Essays on several curious and useful subjects, in speculative and mix'd mathematics.* Illustrated by a variety of examples, J. Nourse, London, 41–51. G2[50]

Sinnott, R.W., 1985. "Astronomical computing—a computer assault on Kepler's equation," *Sky and Telescope,* **70,** 158–159. D

Smiley, C.H., 1934. "Note on Davidson's method of determining the true anomaly and perihelion-time in hyperbolic orbits," *Monthly Notices of the Royal Astronomical Society,* **95,** 63–65. G

Smith, G., 1979. "A simple efficient starting value for the iterative solution of Kepler's equation," *Celestial Mechanics,* **19,** 163–166. D2K[99]

Smith, O.K., 1968. "Terminating the iterative solution of Kepler's equation," *American Rocket Society Journal,* **31,** 1598. D2[99]

Sorokin, N.A., 1982. "About the solution of Kepler equation," *Nachnye Informatsii,* **55,** 135–143. H2

Speiser, A., 1947. "Einteilung der samtlichen Werke Leonhard Euler," *Commentarii Math. Helvetii,* **20,** 288–318. H4

Stephenson, B., 1987. *Kepler's Physical Astronomy, Studies in the History of Mathematics and Physical Sciences,* Springer-Verlag, New York. H3

Stewart, M., 1756. "Solution of Kepler's problem," *Essays and observ. phys. and litt. of the Phil. Soc. of Edinburgh,* **2.** A2K[14]

Struve, O., 1860. "Librorum in Bibliotheca Speculae Pulcovensis Anno 1858 Exeunte Contentorum Catalogus Systematicus," *Acad. Imp. Sci. Petrop.* H1

Stumpff, K., 1968a. "On the application of spinors to the problems of celestial mechanics," *National Aeronautics and Space Administration Technical Note* D-4447. E[96]

Stumpff, K., 1968b. "On the application of Lie-series to the problems of celestial mechanics," *National Aeronautics and Space Administration Technical Note* D-4460. C7[41]

Stumpff, P., 1987. "The general Kepler equation and its solutions," *Celestial Mechanics,* **43,** 211–222. E

Sylvester, J., 1865. "On Lambert's theorem for elliptic motion," *Monthly Notices of the Royal Astronomical Society,* **26,** 27–29.

Sylvester, J., 1866. "On Lambert's theorem," *Philosophical Magazine,* **31,** 52–76.

Taff, L.G., 1985. *Celestial Mechanics: A Computational Guide for the Practitioner,* Wiley-Interscience, New York. H5[33]

Taff, L.G. and Brennan, T.A., 1989. "On solving Kepler's equation," *Space Telescope Science Institute Preprint Series,* 1–15. H2K[107]

Taylor, E.J., 1988. "Solution and iteration surfaces for Kepler's equation," M.S. Paper, Department of Mathematics, Iowa State University. D2K[99]

Tchebychef, P.L., 1857a. "Sur la série de Lagrange," *Bulletin de la Classe Physico-Mathématique de l'Académie des Sciences de St.-Petersburg,* **15,** 289–307. C2[75]

Tchebychef, P.L., 1857b. "Sur la série de Lagrange," *Journal de Mathématiques pures et appliquées (2),* **2,** 166–183. C2[75]

Thiele, T.N., 1883. "Neue Methode zur Berechnung von Doppelsternbahnen," *Astronomische Nachrichten,* **104,** 245–254. H5

Trembley, J., 1782. "Ueber Keplers Problem in Absicht auf die Anweendung desselben in der Astronomie," *Astronomisches Jahrbuch,* 185–187. D1[82]

Udick, S., 1912. "List of errata in Astrand's 'Hulfstablen zur Auflosung der Kepler'chen Problems,'" *Astronomical Journal,* **27,** 97. I

Ueta, Y., 1927. "A mechanical solution of Kepler's equation," *Monthly Notices of the Royal Astronomical Society,* **87,** 207–209. J1[87]

Valson, C.-A., 1868. *La Vie et Les Travaux du Baron Cauchy, Membre de l'Académie des Sciences,* Gauthier-Villars, Paris. H3[68]

Varignon, P., 1705. "Du mouvement des planètes sur les orbes, en y comprenant le mouvement de l'apogée ou de l'aphélie," *Mémoires de l'Académie Royale des Sciences (Paris),* 347–361. H3

Viaro, B., 1920. "Sulla risoluzione della equazione di Keplero," *Atti del Reale Instituto Veneto di Scienze, Lettere ed Arti,* **79,** 965–979. I[84]

Villarceau, Y., 1857. "Détermination des orbites," *Annales de l'Observatoire de Paris,* **3,** 1–147. H2

Voll, W., 1826. "Versuch einer genauen Auflosung des Kepler-schen Problems, wenn die Excentricitat sehr bedeutend ist," *Isis, oder Encyclopadische Zeitung,* **10,** 673–677. H2

von Littrow, J.J., 1815. "Bemerkungen uber Gauss Theoria motus," *Astronomisches Jahrbuch,* 92–100. H6

von Littrow, J.J., 1818. "Anomaliae verae per mediam determinato," *Mémoires de l'Académie Imperiale des Sciences (St. Petersburg),* **6,** 235–255. C2

von Littrow, J.J., 1824. "Ueber die verscheidenen Methoden die Bahn eines Kometen oder Planeten zu bestimmen," *Astronomisches Jahrbuch,* 90–97. H2

Wallace, W., 1835. "Two elementary solutions of Kepler's problem by the angular calculus," *Monthly Notices of the Royal Astronomical Society,* **3,** 127. A1[18]

Wallace, W., 1836. "Two elementary solutions of Kepler's problem by the angular calculus," *Memoirs of the Astronomical Society of London*, **9**, 185–192. A1[82]

Wallis, J., 1695. "Problematis Kepleriani per cycloidem solutio," in *De Cycloide Tractatus, Opera Mathematica*, Volume 1, 540–541. B2[20]

Waterston, J.J., 1850. "On the graphical mode of computing eccentric anomaly," *Monthly Notices of the Royal Astronomical Society*, **10**, 169. B1[21]

Watson, G.N., 1922. *Treatise on Bessel Functions*, Cambridge University Press, Cambridge. H4[31]

Watson, G.N., 1948. *Theory of Bessel Functions*, Second Edition, Cambridge University Press. H4

Weiler, A., 1884. "Uber die Variation der Excentricitat und der Epoche in der gestoren Ellipse," *Astronomische Nachrichten*, **111**, 117–124. H5

Weiler, A., 1899. "Eine neue Losung des Kepler'schen Problems," *Astronomische Nachrichten*, **150**, 247–252. C

Weiss, E., 1881. "Uber die Berechnung der Differentialquotienten der wahren Anomalie und des Radiusvector nach der Excentricitat in stark excentrischen Bahnen," *Sitzungsberichte d. kaiserlichen Akad. d. Wiss., Math-Naturw. Cl.*, **83**, 466–476. C

Weiss, E., 1884. "Entwickelungen zum Lagrange'schen Revertionstheorem und Anwendung derselben auf die Losung der Kepler'schen Gleichung," *Denkschriften der kaiserlichen Akad. d. Wiss. (Wien), Math.-Natw. Cl.*, **49**, 133–170. C2[25]

Whiteside, D.T., 1965. "Newton's early thoughts on planetary motion: a fresh look," *Brit. J. Hist. Sci.*, **2**, 117–137. H3

Whiteside, D.T. (ed.), 1968. *The Mathematical Papers of Isaac Newton*, Volume 2, Cambridge University Press, Cambridge, 206–247. H3

Whiteside, D.T. (ed.), 1974. *The Mathematical Papers of Isaac Newton*, Volume 6, Cambridge University Press, Cambridge, 308–323. H3[20]

Whittaker, E.T. and Watson, G.N., 1927. *A Course of Modern Analysis*, Fourth Edition, Cambridge University Press, Cambridge. H4[29]

Wilczynski, E.J., 1912. "A forgotten theorem of Newton's on planetary motion, and an instrumental solution of Kepler's problem," *Astronomical Journal*, **27**, 155–156. J2[21]

Wilczynski, E.J., 1913. "Note concerning the instrumental solution of Kepler's problem by means of the trochoid," *Astronomical Journal*, **27**, 185. J2[21]

Wilkens, A., 1918. "Zur Auflosung der Keplerschen Gleichung," *Astronomische Nachrichten*, **206**, 21–22. F2[86]

Wilkens, A., 1931. "Die Analogie zum 3. Keplerschen Gesetz bei den parabolischen Kometenbahnen," *Astronomische Nachrichten,* **241,** 9–10. H5

Willis, J.E., 1929. "On facilitating the accurate solution of Kepler's equation by the use of exact equivalents," *Astronomical Journal,* **39,** 117–118. A1[18]

Wilson, C., 1968. "Kepler's derivation of the elliptical path," *Isis,* **59,** 5–25. H3

Wilson, C., 1970. "From Kepler's laws, so-called, to universal gravitation: empirical factors," *Archive for History of Exact Sciences,* **6,** 89–170. H3

Wilson, C., 1989. *Astronomy for Kepler to Newton: Historical Studies,* Various Reprints, London. H3

Wintner, A., 1941. *The Analytical Foundations of Celestial Mechanics,* Princeton University Press, Princeton. H4[38]

Witt, G., 1906. "Zur numerischen Auflosung zweier Gleichungen in der Planetentheorie," *Astronomische Nachrichten,* **172,** 129–138. F2[86]

Wolfers, J.P., 1846. "Auflosung des Kepler'schen Problems nach Newton, verglichen mit der jetzt noch gebrauchlichen numerischen Auflosung," *Archiv für Mathematik und Physik,* **7,** 184–190. H3

Wolfers, J.P., 1849. "Populare Vorlesungen uber wissenschaftliche Gegenstande von F.W. Bessel," *Archiv für Mathematik und Physik,* **13,** 143–159. H5

Wolfers, J.P., 1862a. "Uber das Kepler'sche Problem," *Astronomische Nachrichten,* **57,** 113–120. D1I[82]

Wolfers, J.P., 1862b.   "Aus einem Schreiben des Herrn Professors Dr. Wolfers," *Astronomische Nachrichten,* **57,** 207–208. D1I[82]

Wolfers, J.P., 1863. "Ueber das Kepler'sche Problem," *Bulletin de l'Académie Royale des Sciences, de Lettres, et de Beaux-Arts de Belgique (N.S.),* **15,** 29–46. H3

Wolfers, J.P., 1865. "Ueber die indirecte Auflosung eines Problems, welches bei hyperbolischen Bahnen dem Kepler'schen der elliptischen Bewegung analog ist," *Astronomische Nachrichten,* **64,** 193–196. D4

Wood, H., 1946. "Nomograms for some astronomical computations," *Journal and Proceedings of the Royal Society of New South Wales,* **79,** 153–159. J5[90]

Wood, H., 1950a. "Kepler's problem," *Journal and Proceedings of the Royal Society of New South Wales,* **83,** 150–163. H1[95]

Wood, H., 1950b. "Kepler's problem—the parabolic case," *Journal and Proceedings of the Royal Society of New South Wales,* **83,** 181–194. H1[95]

Zelbr, K., 1895. "Notiz betr. das Kepler'schen Problem," *Astronomische Nachrichten,* **139,** 191–192. B1[21]

Zenger, Ch.-V., 1882a. "Solution rapide du problème de Kepler," *Comptes Rendus (Paris),* **95,** 171–174. A1[18]

Zenger, Ch.-V., 1882b. "Note additionelle sur la solution rapide du problème de Kepler," *Comptes Rendus (Paris),* **95,** 207–208. A1[18]

Zenger, Ch.-V., 1882c. "Tables auxiliares pour calculer l'anomalie vraie des planètes," *Comptes Rendus (Paris),* **95,** 208–210. I

Zenger, Ch.-V., 1882d. "Solution du problème de Kepler pour des excentricités considérables," *Comptes Rendus (Paris),* **95,** 416–417. D4I

Zenger, Ch.-V., 1882e. "On the solution of Kepler's problem," *Monthly Notices of the Royal Astronomical Society,* **42,** 446–449. A1

Index

A

Adams, J.C. 21, 49, 51, 87
Adytum Astronomiae 18
Aitken's method 47
Aldis, W.S. 84, 85
almost units 114
American solution 17
anomaly 1
anomaly, eccentric 2
anomaly, mean 3
anomaly, Mercator's 11, 12
anomaly, Seth Ward's 9, 13
anomaly, true 2
Astrand, J.J. 82, 84
Astronomische Nachrichten 80
autonomous differential equations 41,
      135

B

Bailey, M.E. 79
Barker, T. 57, 58
Barker's equation 57, 58, 84, 142, 145
Barnsley, M., *Fractals Everywhere* 101
Battin, R.H. 101
Bergam, M. 101, 105
Bessel, F. W. 27, 28, 29, 30, 53, 60, 61,
      65, 73, 84
Bessel functions 23, 24, 27, 28, 29, 30,
      33, 36, 37, 116
Bessel's interpolation formula 84
Bina, S. 113
Bobone, J. 94
Boltz, F. 103, 104, 147
Boulliau, Ismael 9, 11
Boulliau's correction of Ward's
      hypothesis 9
Boulliau's hypothesis 9
Bourget, J. 34, 35, 73
Bowditch, N. 25, 59, 61
Brenke, W.C. 7, 8
Brennan, T. 107

C

Brinkley, J. 9, 10, 11, 12, 13, 14, 47, 53
Broucke, R. 100, 101, 102, 103, 104
Brunnow, F. 65
Bryant, R. 25
Burkhardt, T. 102
Bürmann, H. 24
Burniston, E. 109, 111
Burniston-Siewert method 108, 109,
      149, 151, 152

Cajori, F. 48, 49, 50, 51
Callendreau, O. 88
Carlini, F. 15, 16, 32, 33, 34, 73, 87
Cassini, Jacques 12, 13, 14, 18
Cassini, Jean-Dominique 11
Cassini solution 12, 13, 14, 52, 53, 87
Cauchy, A. 23, 34, 67, 68, 69, 70, 71, 72,
      73, 74, 75, 76, 112, 116, 151
Cauchy numbers 73
Cayley, A. 34
Comes Paganus 9
Conway, B. 104, 105, 106
Creedy, C. 25
Curtz, A. 8, 10, 11
curve of sines solution 21, 87
cycloid 20, 87
cycloid solution 20

D

Danby, J. 102, 104, 105, 106
Dazew, A. 88
DeGasparis, A. 81, 82
Degen, C.F. 25
De la Caille, N.L. 46
Delves-Lyness method 108, 112, 113
Doberck, W. 82
D'Ocagne, M. 91
Dolainski, F. 87
Draper, G. 85
Dubois, E. 21, 87

E

Encke, J.F. 14, 15, 53, 54, 57, 82
Englefield, H., *Orbits of Comets* 58
Euler, Leonhard 19, 23, 29, 46, 50, 67, 82

F

fixed point 45
Fleckenstein, J.O. 37, 117
formal power series 113
Fourier, J. 51
Fry, T.C. 91

G

Gauss, C.F. 30, 48, 50, 61, 64, 65, 66, 67
Gauss' method 16, 61, 66, 82
Gaythorpe, S.B. 10, 16, 17
Glaisher, J.W.L. 35
Gonggrijp, B. 21
Gooding, R.H. 102, 104, 105, 106
Greathead, S. 34
Grobner, W., *Die Lie-Reihen und ihre Anwendungen* 41
Grunert, J.A. 17, 18, 47
Gylden, H. 35

H

Habash-al-Hasib 4, 5, 46
Hagihara, Y. 37
Hall, A. 32
Halley, E. 26, 57, 58, 60, 61, 65
Hansen, P.A. 33
Hennert, J.F. 31
Herget, P., *The Computation of Orbits* 66
Herrick, S. 94
Herz, N. 83
Hilbert boundary value problem 109, 151, 152
Horrebow, Peder 18, 19
Horrocks, Jeremiah 10, 11, 16, 18, 19
Horrocks solution 16, 18, 121
Howe, H.A. 17
Hurwitz, A. 76

I

Innes, R.T.A. 94
Ioakimidis-Papadakis method 108, 111
iterative sequence 45
Iushkevich, A.P. 5
Ivory, J. 54, 55, 56, 60

J

Jacobi, C.G.J. 33
Jekhowsky, M.B. 36
Jorgensen, N.T. 18, 19

K

Kapteyn, M.W. 36
Kapteyn inequalities 39
Kapteyn series 36
Keill, D.J. 51, 52
Kennedy, E.S. 5
Kepler, Johannes 1
Kepler problem ix
Kepler problem, bibliographies for x
Kepler's equation ix
Kepler's equation, Battin's universal 141, 145
Kepler's equation, multivariable 36, 37
Kepler's equation, universal 96
Kepler's graphs 89
Kepler's laws 1
Kepler's solution 4, 46, 104
Klugel, G. 25

L

Lacroix, S.-F. 50
Lagrange, J.L. 24, 27, 29, 31, 43, 68, 69, 115, 117, 122, 127
Lagrange-Bürmann expansion theorem 24, 26
Lagrange's theorem 23, 25, 26, 29, 32, 68, 69, 72, 75
Laguerre's method 104, 106
Langebartel, R. 94, 95
Laplace, S.P. 25, 26, 27, 50, 59, 68, 69, 72, 74, 79
Laska, V. 91
Laurent's theorem 33, 68
Legendre, A.-M. 50, 72
Le Grand Roy, E. 91
Lehmann, W. 80, 84
Levi-Civita, T. 23, 38, 43, 115, 129, 132, 133
Levin, A.E. 88
Lie-series 41, 136, 137, 157
logarithmic derivative 75

M

Mache, H. 88
Machin, J. 52, 53, 125, 126
Marth, A. 83
Mécanique céleste 25, 26, 59, 68, 72
Mercator, Nicolas 11
Mikkola, S. 106, 107
Moller, J.P. 94
Moulton, F.R. 26, 43, 76, 87, 94, 99
Mouraille, J. 50

N

Newton, Isaac 20, 49, 50, 51, 53, 67
Newton's identities 113
Newton's method 15, 48, 49, 52, 53, 82,
        89, 90, 95, 98, 99, 100, 101,
        102, 104, 107, 113, 115, 126,
        127, 132, 138, 155
Newton's method for power series 108,
        113, 114, 116, 129, 155
Newton-Raphson method 48, 49
New York Mathematical Society 36
Ng, E. 99, 100, 104
Niele, William 9
Nijenhuis, A. 107
nomograph 89

O

Odell, A. 102, 105
Olbers, W. 30
Oppolzer, Th. v. 48, 66, 83
Osgood, W.F. 69

P

Palitzsch, J. 58
parabolic approximation 56, 57
parallax, equitorial horizontal 120
parallax, geocentric 119, 120
parallax problem 4, 119
Paul, E. 43
Peters, C.F.W. 47, 48
Peters, R. 102, 103, 104
Picard, E. 76
Pitkin, E. 95
Plummer, H.C. 17, 21, 88
Poisson, S.D. 31, 32
Pritchard, C. 87
Privaloff boundary value problem 109,
        150, 151, 152
Proctor, R.A. 87

Prussing, J. 100, 101, 105
Puiseux, V. 72, 74, 75, 76

R

Radau, R. 90, 91
Rambaut, R.A. 21, 88
Raphson, J. 50
reversion of series 24
Riemann boundary value problem 109,
        149
Robertson, A. 52
Rouché, E. 75

S

Scheibner, W. 33
Schröder's method 100
Schubert, E. 53, 98, 102, 106, 107
Schubert, F. T. 25, 26, 128
Sconzo, P. 32, 53
Searle, G.M. 66
See, T.J.J. 87
Serafin, R. 107
Serret, J.-A. 72, 75, 76
Seydler, A. 82
Sheela, B. 101, 103, 104
Siewert, C. 109, 111
Simpson, T. 50, 59, 60, 65
Smith, G. 99, 100, 103
Smith, O.K. 99
strange attractors 100
Stumpff, K. 41, 43, 116, 136, 157

T

Taff, L.G. 80, 107
Taylor, E. 99
Tchebycheff, P.L. 75
Tchebycheff's formula 100
Trembley, J. 82

U

Ueta, Y. 87

V

Viaro, B. 84

W

Wallace, W. 82
Wallis, J. 20, 50
Ward, Seth 9
Waterston, J.J. 21, 87

Wegstein, J. 108
Weiss, E. 25
Whiteside, D.T. 48, 49, 50, 51
Wilczynski, E.J. 21, 88
Wintner, A. 39, 69, 76
Wolfers, J.P. 82
Wood, H. 95
Wren, Christopher 20, 51

Z

Zelbr, K. 87